Studies in Systems, Decision and Control

Volume 519

The series "Studies in Systems, Decision and Control" (SSDC) covers both new developments and advances, as well as the state of the art, in the various areas of broadly perceived systems, decision making and control–quickly, up to date and with a high quality. The intent is to cover the theory, applications, and perspectives on the state of the art and future developments relevant to systems, decision making, control, complex processes and related areas, as embedded in the fields of engineering, computer science, physics, economics, social and life sciences, as well as the paradigms and methodologies behind them. The series contains monographs, textbooks, lecture notes and edited volumes in systems, decision making and control spanning the areas of Cyber-Physical Systems, Autonomous Systems, Sensor Networks, Control Systems, Energy Systems, Automotive Systems, Biological Systems, Vehicular Networking and Connected Vehicles, Aerospace Systems, Automation, Manufacturing, Smart Grids, Nonlinear Systems, Power Systems, Robotics, Social Systems, Economic Systems and other. Of particular value to both the contributors and the readership are the short publication timeframe and the world-wide distribution and exposure which enable both a wide and rapid dissemination of research output.

Indexed by SCOPUS, DBLP, WTI Frankfurt eG, zbMATH, SCImago.

All books published in the series are submitted for consideration in Web of Science.

Domenico Marino · Daniele Cananzi ·
Filippo Aragona

Ethics and Artificial Intelligence

Towards a Moral Technology

Springer

Domenico Marino
Department of
Heritage-Architecture-Urbanism (PAU)
Mediterranea University of Reggio Calabria
Reggio Calabria, Italy

Daniele Cananzi
Department of Law, Economics and Human Sciences (DIGIES)
Mediterranea University of Reggio Calabria
Reggio Calabria, Italy

Filippo Aragona
Italian Ministry of Justice
Reggio Calabria, Italy

ISSN 2198-4182 ISSN 2198-4190 (electronic)
Studies in Systems, Decision and Control
ISBN 978-3-031-50901-8 ISBN 978-3-031-50902-5 (eBook)
https://doi.org/10.1007/978-3-031-50902-5

This Springer imprint is published by the registered company Springer Nature Switzerland AG
The registered company address is: Gewerbestrasse 11, 6330 Cham, Switzerland

Introduction

In the era in which artificial intelligence emerges as one of the driving forces behind social and technological transformations, ethical questions arise that recall the very essence of our humanity and invite us to reflect deeply on our relationship with the machines we have created. The book Ethics and Artificial Intelligence proposes to explore in a thorough and articulate manner the complex and delicate intertwining between the evolution of thinking machines and the moral values that should guide humanity's choices in this evolving scenario. Artificial intelligence, understood as the ability of machines to learn, process information and make human-like decisions, has reached astonishing levels of complexity and versatility. Its applications range from more precise medical diagnoses to autonomous driving, from predicting weather patterns to optimising supply chains. These advances arouse wonder and hope, but at the same time raise profound questions about the effects AI could have on society, the economy and the individual. One of the most central ethical questions concerns the autonomy of machines and the concept of responsibility. As AI comes closer and closer to being able to make complex decisions autonomously, the dilemma arises as to who is responsible when AI decisions lead to negative consequences. For example, if an autonomous vehicle causes a fatal accident, who is at fault: the programmer, the vehicle owner, the company that developed the algorithm? The answer to this question has profound legal, moral and social implications, requiring a rethinking of existing legal and regulatory structures. Privacy is another area where AI raises complex ethical issues. With the massive collection of data and the ability to analyse it to draw surprising inferences, concerns arise about constant surveillance and the potential violation of the individual sphere. The balance between technological innovation and the protection of individual rights is a crucial issue to explore, as decisions on how to use data collected by AI can have significant impacts on people's freedom and self-determination. Technological unemployment is another AI-related ethical challenge. As machines become capable of performing tasks previously reserved for humans, fears arise that many professions may be automated, leading to job losses and growing economic inequality. However, it is important to recognise that AI can also create new job opportunities, especially in the design, maintenance and development of these technologies. The ethics of AI are also intertwined with the challenges of equity and social justice. Machine learning algorithms can reflect unconscious biases in training data, leading to discriminatory decisions in areas such as recruitment, funding and access to essential services. It is crucial to address these issues proactively by adopting ethical design practices that take diversity and inclusion into account. In addition to practical challenges, AI also raises philosophical and conceptual questions that invite us to explore the very nature of intelligence and consciousness. What does it mean to be intelligent? Do the algorithms that emulate human intelligence really have an understanding or consciousness of the world, or are they simply performing complex mathematical operations? These questions prompt us to reflect on the distinction between artificial intelligence and human intelligence, paving the way for philosophical debates on the capabilities and limitations of machines compared to human creativity, empathy and understanding. The book is meant to be a stimulus to carefully consider how we want to shape the future of AI in a way that reflects our core human values. The challenge is to take a responsible and thoughtful approach to technological innovation so that we can make the most of AI without compromising what makes us truly human. Chapter Contribution—In the context of attributions, it is essential to highlight the contributions of specific individuals to the development of this volume. Chapter 1 was contributed by Filippo Aragona, Chap. 2 by Daniele Cananzi, and Chaps. 3 and 4 by Domenico Marino. While these individual contributions are pivotal, it must be emphasized that the creation of this volume has been the culmination of an intense group effort.

Contents

Chapter 1
The Fil Rouge Between Human Beings and Artificial Intelligence

Domenico Marino, Daniele Cananzi, and Filippo Aragona

Trasumanar significar per verba non si poria
(Divine Comedy, Paradise, Canto I)

1.1 A Brief Account of the Human Philosophical Thought Linked to Artificial Intelligence

If one were to imagine a distant, very distant future in which one were to ask a being endowed with artificial intelligence what man is, the answer might be conditioned by the reverential awe that one generally attaches to a parent or progenitor, and thus that form of intelligence would probably refrain from defining him as man has defined himself, i.e. as the world's greatest and most disturbing enigma, who can know himself from below and above, from his own light (id est from the divine principle within him) or from his darkness (id est from the demonic principle), a dual and contradictory being, polarised to the supreme degree, god-like and beast-like, capable of elevation and lowering, of great love and sacrifice but also of great cruelty and unlimited selfishness, to such an extent that no man is so evil as to be incapable of goodness, just as no man is so good as to be incapable of evil, because in the holy there is the power of the criminal and in this that of the holy.[1]

If, then, that artificial being endowed with intelligence were capable of self-critical analysis, he might even be tormented by something unbearable, thinking of the fascination that comes from man's duality and ambiguity and the irrationality and contradictions that make him so perturbing and destabilising. The intelligent artefact, rather than savouring the pleasure of its ancient origins dating back at least as far as Leonardo da Vinci, would experience moments of disorientation as it recalls ancestors capable of violent passions, burning lusts, fire in the eyes, premonitory dreams, feelings and

[1] Nikolaj Berdjaev, Sophocles.

D. Marino et al., *Ethics and Artificial Intelligence*, Studies in Systems, Decision and Control 519, https://doi.org/10.1007/978-3-031-50902-5_1

presentiments, creative ferment, poetic synaesthesia and musical harmonies between eros and thanatos.

And if that intelligent object were able to connect the abstract and universal concepts of physics and metaphysics, it would be dumbfounded by the fact that man is not a pre-packaged artifice programmed by others but is a natural product of chance and necessity.

But the question that would worry him the most is why on earth man at some point in his evolutionary history decided to spend most of his energy on creating his own simulacrum, namely him, the object with artificial intelligence.

Attempting to answer this question could lead him through the history of human thought to trace the common thread that holds his existence together.

He would begin his research in the field of the *humanities* before that of science, for it is in that field that the abysses of madness were first plumbed, among whose chiaroscuri the most disturbing being in the universe has evolved over the millennia.

After having run the risk of getting lost in the sensual and intoxicating labyrinths of poetry, tragedy and Greek philosophy, the system of electric circuits endowed with intelligence would dwell at length, probably in a state of *dizziness* and *awkwardness*, on Cartesian dualism between *res cogitans* and *res extensa* and on Cartesian doubt as the starting point of all scepticism, especially with reference to the formation of self-consciousness. He would discover that from that doubt modern philosophy was born and the main conceptions that paved the way for contemporary thought branched out: rationalism and empiricism, dogmatism and criticism, idealism and realism, rational philosophy and irrational philosophy. He would also discover that that doubt is the main causal antecedent of his existence and the prodrome of his destiny, because from it arose the knot of the *brain-mind* relationship that has involved philosophers and scientists and that has never been resolved, since the opposition between those who see in *consciousness* the most mysterious and unresolved side of human nature and the reductionist thinkers who bring that binomial back to unity, believing that *brain* and *mind* are the same thing, has not been overcome.

The intelligent object would attempt to get closer to the knowledge of its origins by following a further path, namely that of the 'technoprogressive' evolutionary conception, according to which human evolution (unlike the Darwinian evolution advocated by 'bioconservative' thinkers) can undergo acceleration through the use of technology, which can lead to posthumanism after a phase of transhumanism. It would thus be realised that technology would not only be used to create artificial intelligence, but also to bring about the enhancement of man's physical and mental faculties to the point of making him mutate into posthuman, i.e. a mixtum of cells and algorithms that would still make him fall into the genus of artificial intelligence or at any rate place him in a genus that is no longer human.

However, this latter path could complicate matters, because from it would emerge infinite variables to decipher that would add new levels of complexity to the search. The intelligent thing could then get lost in the secret webs of life and end up inside a labyrinth of mirrors that return increasingly deceptive and misleading images until it is forced to begin its journey again from the beginning and then find itself faced with continuous bifurcations.

Considering that there is an infinitesimal chance that this book you are reading will arouse the interest, who knows how many centuries from now, of an artificial intelligence curious to discover ancient writings, we could try to give some pointers to the intelligent object that would like to delve into the search for its origins, helping it at least to better understand certain passages in the history of thought that have pushed man to create his alter ego either through artificial intelligence or in pursuit of the dream of posthumanism.

First of all, before focusing on the consequences of doubt and Cartesian dualism, it would be useful to dwell for a moment on the thought of Fuerbach, who most probably and unwittingly identified the *punctum dolens of* this whole affair.

In fact, the German philosopher examined man's need to have limits with which to clash and confront in order to create a space–time dimension that can be dominated by his intellectual capacities and to discover his own essence. The first limit is given first of all by the object with which man's activity is measured and which reveals the human being and his nature 'alienated' in the object. Therefore, when one becomes conscious of an object, one also acquires awareness of one's own being because the object becomes the self-assertion of man who grasps himself in that object, an expression of himself outside himself.

According to Feuerbach, even 'god is but an expression of us outside ourselves, in which our own nature, characterised by thought, will, and heart, is revealed, objectivised. The force of thought is the light of knowledge, the force of will is the energy of character, the force of heart is love'.[2] So man created god by attributing to him these human qualities, namely reason, will and love, alienating himself into god. Therefore god is the derivative of man. God is the predicate, while man is the subject: deus imago hominis. With the creation of god, however, man has also constructed a limit that is erected against him; in fact, by alienating himself in god, man has also subjected himself to god, renouncing to live spontaneously according to the natural rules of becoming and accepting an artificial system built on the alternation of rewards and punishments as the consequences of predetermined and pre-packaged rules whose sole purpose is to keep intact the erected limit to which the name of god has been attributed.

The need to enclose oneself within narrow but reassuring limits has always accompanied man throughout his history; in fact, this need gave rise to Socratic philosophy and then later philosophy up to German idealism. The whole course of ancient and modern philosophy is in fact characterised by the construction of epistemic, incontrovertible, absolute realities, which, on the one hand, should have satisfied man's desire to know what is hidden behind the veil of Maya, but, on the other hand, denied the evidence of the πάντα ῥεῖ Heraclitean and the unpredictability that distinguishes it. All philosophical (and religious) constructions have attempted to make sense of life by considering it an expression of an immutable law of the whole and thus rejecting the terrifying idea that life can be born from nothing, become and then end in nothingness.

---

[2] Mario Dal Pra, Sommario di Storia dela Filosofia (Volume III) La Nuova Italia, Firenze, 1984.

Already this first reflection could give rise in the intelligent object to the idea that he would be the new limit that man has erected and into which he has alienated himself after attributing his own qualities. Having broken down the concept of god along with all epistemic conceptions, man may have felt the need for a new god, closer to him and usable ad libitum.

On the other hand, contemporary philosophy, initiated above all by Artur Schopenauer, Giacomo Leopardi, Auguste Comte, Saint Simon, John Stuart Mill, Jeremy Bentham, and Friedrich Nietzsche, has demolished not only the concept of god, but also all the rigid rational or religious scaffolding and superstructures that stifle the becoming and the wonder of the unpredictability and authenticity of things, inciting man to say an unconditional 'yes' to life and its continuous metamorphoses and infinite possibilities, boldly looking at reality in both its horrendous and fascinating aspects, without hiding behind rigid artificial schemes programmed by thought not to see in order not to suffer. The remedy that had been chosen by traditional philosophy, however, proved to be worse than the evil, because in order to live without facing the unpredictability of life, enclosing it within sophisticated epistemic and religious constructions in order to shelter oneself from the idea that reality comes from nothing and ends in nothingness and to anaesthetise the thought of death, true life has been killed off, i.e. the life characterised by imponderable changes, randomness and death as the antipode in which life must necessarily end if it is to have any meaning (each element only acquires an identity if it is united with its opposite, otherwise it would be emptied of its essence and implode into itself).

The process of demolition and removal of all immutable and absolute truth determines the liberation of modern man from the heavy philosophical and religious chains that prevent him from fully experiencing all aspects of existence, including those characterised by disorder and chaos.

With contemporary philosophy, then, scepticism returns, which, however, should not be (between) understood as a form of thought that denies any truth, because it actually consists in the assertion that there is not or cannot be any form of epistemic truth that stands above the truth of the world of becoming. This conception represents the link between Nietzsche's philosophical thought and that of Heraclitus, in that for both the only truth is the incessant becoming of things that cannot be denied by epistemic constructions of reality.

In this context also arises the positivism inaugurated by Auguste Comte, according to which social and human progress can only be achieved by overcoming the primitive phase of myth and theology and the intermediate phase of metaphysics, with the consequent abandonment of the search for the first and last causes of reality, imprisoned in an immobilising framework of epistemic truths, in order to be able to devote oneself instead to understanding phenomena as they occur in the mutations of becoming.

Man's liberation from the constructions built to protect himself from the unpredictability of becoming reaches its climax with Nietzsche, whose revolutionary vis can be grasped in the words with which he described the power of philosophical thought in his third "Inactual Consideration, Schopenhauer as Educator": "an American can tell them what a great thinker coming to this earth means, as a new centre of enormous forces. Be careful, says Emerson, when the great God sends a thinker to our planet. Everything is then in danger. It is like when a fire breaks out in a big city and no one knows what is really safe and how it will end. Then there is nothing in science that cannot be overthrown in the aftermath, then no literary prestige, no so-called eternal celebrity is worth any more; all the things that are dear and precious to man at this hour are only so by virtue of the ideas that have risen to his spiritual horizon and that have produced the present order of things as a tree bears its apples. A new degree of culture would instantly subvert the whole system of human aspirations.

The immense spaces left open by contemporary philosophy and the absence of traditional points of reference and orientation have evidently induced man to take refuge within new schemes that have found their development in artificial intelligence, through the realisation of something predeterminable, programmable and controllable that could shelter him from the changes of events and the terror of nothingness, without however considering the risks of degeneration that could result. In fact, this choice to (re)tread reassuring paths, paved with repetitiveness, determinism, calculability, contains a deflagrating destructive power, because the most fascinating traits of life are anaesthetised: chance, spontaneity, freedom, irrationality, contradiction, paradox, creativity, the unknown and the forbidden.

The path indicated by Nietzsche to overcome mankind went in another direction, but was not followed perhaps because it was too vertiginous: 'those who know how to breathe the air of my writings know that it is an air of the peaks, a strong air. One must be born to breathe that air, otherwise one runs the risk, not a small one, of getting cold up there. The ice is near, the solitude immense… Philosophy, as I have understood it and lived it to this day, is voluntary life among the ice and the heights, a search for everything that existence has that is foreign and problematic…'.[3] Nietzsche had in fact imagined that the human being could overcome himself without any external intervention, managing to walk the tightrope over the abyss thanks only to his own ability to face life in all its aspects, both Apollonian and Dionysian. The superman is in fact the one who can experience all the forms arising from chance without any fear of the multiple sides that cannot be predicted. He is thus an expression of the widest freedom and will to power, two factors that imprint becoming with the character of being and give a sense of eternity. This relates to the conception of the 'I' in Fichte's thought, according to which the 'I' has the infinite task of removing that which is absolutely opposed to it, i.e. it has the infinite task of becoming 'god'. The allegorical forms of the Prometheus and Faust come close to this conception of the 'I', who unleash their genuine freedoms and capacities in an authentic manner (i.e. without artificial grafts deriving from *ante litteram* technologies) up to the extreme courageous attempt to approach the truth, the one through the theft of fire from the gods (fire to be understood as the *arché* of all things and therefore as absolute truth)

[3] Ecce Homo, Nietzsche.

and the other through a pact with Satan in order to know what passes through all things, generating them and keeping them united in the eternal logos.[4]

On the other hand, the history of man's philosophy and thought stems from the intertwining of *mythos* and *logos*, that is to say, of feeling and reason, and the entire subsequent period flows along the labyrinth traced by this link, whose roots lie in the uninterrupted murmur of being in the heart of the cosmos.

Going back to Nietzsche, it could be said that the superman is the one who discovers the synthesis between mythos and logos by immediately experiencing mysteries, visions, enlightenments, thunderbolts, thoughts arising from revelations, enigmas. The superman makes thought flow into the unceasing becoming of the cosmos, rather than using it to erect artificial constructions to contain the pressures arising from the universe, having the courage to look life in the face with all its wonders and horrors and its disquieting unpredictability and randomness. He thus frees himself from rational and logical scaffolding that, from Socrates onwards, have anaesthetised man by deluding him into believing that there is an immutable reality from which everything comes and into which everything returns. The rigid structures created by rational philosophy, including the figure of god, have ended up erasing the authenticity of man's life, inducing him to live within the narrow limits set by reason and logic and preventing him from doing so in a genuine and authentic way (if god exists, man cannot live). The gigantic edifice—culminating in the concept of god—built by western culture and civilisation to protect man from chaos and randomness has ended up burdening man's existence with a weight far more unbearable than that constituted by the very fear of becoming and chance. The pleasure of insecurity is proper to the superman who does not recoil and does not try to protect himself from life's changes, but accepts life unconditionally even in its most terrible, darkest and harshest aspects. The superman, completely disenchanted with all the illusory remedies created by rational philosophy and religions, does not flee in the face of becoming but identifies with it.[5]

The revolution imagined by Nietzsche would have exalted the genuineness of life by bringing to light the immediacy of intuition in the folds of becoming. It would have been a genuine revolution, as 'what is essential for a revolutionary is not the overthrow as such, but the fact that, in the overthrow, he brings to light what is decisive and essential'.[6]

Instead, what has been thought of as a revolution (the technological one, based above all on Artificial Intelligence) would appear to be something quite different, because on the one hand it contains germs that are destructive of mankind to the point of obscuring millennia-old evolutionary processes, and on the other hand it re-proposes the outdated logical schemes that stiffen life within pre-constituted, programmed and predictable bottlenecks. Heidegger had pointed out this risk, in fact he had seen in 'technique' the trap that would make the immutable realities

[4] Transhumanism and Posthumanism, Domenico Marino and Filippo Aragona, Digital Agenda, 2019.

[5] Filippo Aragona, Mythos—Logos, GNOSIS Magazine, 3/2019.

[6] Nietzsche, by Martin Heidegger.

distant from reality that founded metaphysics re-emerge: "precisely technique is the phenomenon that expresses, on the level of man's way of being in the world, the unfolding and coming to fulfilment of metaphysics. To the coming into being only more as will...corresponds the modern technique that gives the world that form we now call total organisation. The metaphysical systems of the nineteenth century are a form of total organisation at a level that is still theoretical...the total organisation implemented by technology is no longer only in theory but is actually realised as the order of the world... the being of the entity is reduced to technical instrumentality...".[7]

However, it has always been imperfection that has allowed continuous mutations of the marvellous mechanism that is the human brain. Machines, despite their instrumental value in achieving 'total organisation', will never be able to mark an evolutionary path like that of humans.

1.2 Origin of the Artificial Intelligence

In order to give an answer to the question posed by our intelligent object on the origins and purpose of its existence, further reflections on the origins of Artificial Intelligence, Transhumanism and Posthumanism are required, bearing in mind that at the basis of the distinction between a robot and a posthuman (composed of electronic artifacts and algorithms grafted into the human body) there is always the gap between artefact and natural element, in relation to which Jacque Monod, in his famous essay 'Chance and Necessity', identified teleonomy and autonomous morphogenesis as the main factors separating artificial objects from natural ones, on the understanding that our intelligent object reading this book could find itself at a point in history where even that dichotomy is absorbed into a *tertium genus* that is no longer either robot or posthuman.

Teleonomy is a fundamental property of all living beings, namely 'that of being endowed with a project, represented in their structures and at the same time realised through their performances'[8] which can also consist in the creation of artefacts. While autonomous morphogenesis consists in the fact that living beings 'owe practically nothing to the action of external forces, while they owe everything, from the general form down to the smallest detail, to internal morphogenetic interactions'.[9]

Artificial Intelligence' originated at a seminar organised in 1956 in New Hampshire by Mc Carthy [1], M. Minsky, N. Rochester and C. Shannon in order to initiate avenues of study to enable machines to do what, if done by humans, would require the use of intelligence. Thus began a new multidisciplinary strand of research in which scientific, philosophical and technological disciplines converged.

---

[7] Introduction to Heidegger by Gianni Vattimo, Laterza Editori, 2000.

[8] Chance and necessity, Jacques Monod.

[9] Chance and necessity, Jacques Monod.

In the attempt to reproduce behaviours that are indistinguishable from human ones, one has come up against an insurmountable limitation, namely the possibility of making the computer assimilate the cultural substratum based on preconceptions, beliefs and customs that are thousands of years old and around which the human personality has evolved. Many have therefore preferred to fall back on a weaker conception of Artificial Intelligence, accepting a more recessive role of the computer, at the service of man and not in competition with man (hence the effective French term for the computer: ordinateur). On the other hand, one must consider that the objectives for which the robot acts are always identified by man, who programmes it for certain purposes that are unavailable to the computer.

One of the main arguments of the proponents of the weak conception is that artificial intelligence is not a true intelligence because it lacks the motivational complex that constitutes the drive for human actions and is divorced from the daily experience and suffering of human beings interacting in a natural and social environment that generates fear, suffering, melancholy, anguish, terror, uncertainty, doubt, crisis, anxiety, depression, madness, grief, nostalgia, euphoria, joy, excitement, enchantment, dreams, fantasies, passions, emotions, delusions, ecstasies, disturbances, artistic creativity, ennui, spleen, horror, poetic and musical intuition.

By now, more and more researchers have abandoned the concept of strong Artificial Intelligence and the illusion of having found in computers the key to a synthetic mind capable of acting as a replica of our brain. In fact, most research is now geared towards the invention of useful machines, i.e. devices capable of acting as effective tools especially for performing social, work and organisational tasks.

To overcome some of the difficulties in which the supporters of the strong conception of Artificial Intelligence found themselves, recourse was made to the functionalist paradigm (adopted in the philosophy of the mind by Putnam's doctrine [2], according to which mental events are qualified by functions, that is, by tasks and roles directed towards a purpose without a specific material substratum) or to the connectionist paradigm (which is based on the conception of the brain as a metaphor of the mind, for which it is possible to reproduce the mind and intelligence through their simulation by means of neural networks). Through these new avenues, an attempt has been made to re-propose an Artificial Intelligence capable of simulating all human behaviour and human thought (an anthropomorphic model, which differs from the one that aims only to emulate, rather than simulate, human intelligence), relying above all on the equation according to which the mind is to the brain as software is to hardware, with the consequence that the mind would be like a programme that can operate in different physical materials, biological-neuronal in the case of the brain or electronic in the case of computers, thus endowing any type of element existing in *rerum natura* with intelligence and thought. This conception is based on the distinction between brain and mind, and recalls all the fascinating theories on consciousness and pamphyschism that will be discussed later, because it is precisely from these that an answer to the question posed by the intelligent object could be found. According to the materialistic view, however, *brain and mind are one and the same* thing, and *consciousness* would depend on the neuronal processes that govern physical actions. Furthermore, while in a computer the software and hardware are separate, in humans brain and mind, according to materialistic thinking, would form a 'wetware'.

A third conception aims at a cooperation between Artificial Intelligence and human intelligence, which are thus placed neither on a competitive nor on a subordinate level. They should converge towards the same end: man's transition towards posthumanism.

The relationship between the two forms of intelligence is also relevant to the issue of what computers can or cannot do or must or must not do and the consequences in terms of social relations, rights, duties and responsibilities.

If the computer is a simple ordinateur, then it will always be the human user who will be blamed for the harmful consequences of the computer's behaviour (according to the well-known *ubi commoda eius et incommoda* principle), whereas if it were really possible to create an Artificial Intelligence that replicates the human one or at least cooperates with the human one, then, within the limits in which it is recognised as having decisional freedom, the Artificial Intelligence could be considered an autonomous subject with rights and duties. A problem of indemnity capacity will arise, but this can be overcome with ad hoc insurance contracts. Certainly, however, their protection from violent, discriminatory and prejudicial acts will have to be considered. The issue is particularly relevant because it is connected to that of the so-called non-human persons, i.e. those beings who, although they are not human, are nonetheless endowed with an intellectual and emotional sphere very similar to the human one and who should therefore be recognised as subjects of law, at least in terms of their capacity to be bearers of interests worthy of protection by the legal system. Emblematic in this sense is the recent case of the two chimpanzees (Hercules and Leo) temporarily imprisoned and separated from their group on Long Island in the USA in 2014 because they were too lively and annoying. The organisation Nonhuman Rights Project initiated a court action that went all the way to the Supreme Court to ensure that Hercules and Leo would be granted the guarantees of *habeas corpus*, believing that the two non-human beings were capable of suffering the same torments as humans, were able to remember past events and plan future actions, to perceive pain and suffering, and to act not only on the basis of instincts but also on the basis of autonomous self-determination.

The US courts, however, after addressing the issue of what should be meant by 'person' before the law, especially with reference to fundamental freedoms, held that the chimpanzees could not be considered subjects of law like human persons, since their intellectual limitations prevent them from being considered holders of the rights and duties that presuppose freedom of choice and the resulting responsibilities.

Mutatis mutandis, the same issues will soon arise with regard to robots,[10] which, although currently lacking the depth and expansive visage of the human soul and thought, could be considered 'non-human persons' in the same way as has been argued for chimpanzees.

---

[10] The term 'robot' was coined in 1920 by Karel Capek from the Czech word 'robota' meaning 'forced labour'.

In the case of robots, there are also serious issues of liability allocation in the event of harmful acts against third parties, as well as ethical issues of paramount importance, especially with reference to those programmed for destructive purposes, such as robot soldiers. It will be recalled the heartfelt appeal to the UN in 2017 by the most important robotics and artificial intelligence companies asking the UN to intervene to ban the application of new technologies to build combat robots in order to avoid the devastating effects that would result ('once Pandora's box is opened, it will be difficult to close it again'). This call had been preceded by the warning of numerous representatives of the world of scientific research meeting in Buenos Aires in 2015, who had highlighted the very high risks for mankind arising from the application of Artificial Intelligence to the field of armed conflict. At a forum in Paris in 2019, the UN Secretary General called on all nations to conclude an international treaty to ban the production of such devices capable of autonomously selecting targets to strike ('Machines that have the power and discretion to kill without human intervention are politically unacceptable and morally despicable'). He also emphasised the contradiction between 'technological progress' and 'regression in human rights', adding that "we should instead be ensuring that artificial intelligence is used to guarantee that everyone can live in dignity, peace and prosperity".

1.3 The Relationship Between Brain and Mind and the Hard Problem of Consciousness

Returning to the relationship between brain and mind, which underpins the theoretical elaboration of Artificial Intelligence, Transhumanism and Posthumanism, it is based on the Cartesian dualism between *res extensa* (brain) and *res cogitans* (mind) and the subsisting relationships between the knowing subject and known reality. Descartes considered mind to be an immaterial substance that interacts with the body (of which the brain is a part) and guarantees, on the one hand, the freedom and autonomy of the inner subject from the mechanistic and deterministic universe (*res extensa*) and, on the other hand, the existence of self-consciousness. He inaugurated the thought of interactionist dualism, which was the main point of reference for all subsequent philosophy, at least until German idealism.

After the dualistic problem posed by Descartes, the concept of *conscience was* joined by that of *consciousness*: the former mainly concerned the moral sphere, while the latter introduced the psychological question into philosophical studies. In fact, even before Descartes, it was Thomas Campanella who had deepened the study of man's innermost sphere (but Augustine and his Neo-Platonic school had already focused on man's inner life), dwelling on his most irrational side and marking the start of the long road that led to modern psychoanalysis.

Tommaso Campanella developed the doctrine of self-consciousness from the premise that 'the soul is known only with news of presence, not as an object, only when it reflects on itself; that we are and can, know and will is the first certain principle…The soul is alienated from news of itself by means of objects, and objects do not manifest themselves either wholly or distinctly, but only in part and in a confused manner'. From these considerations Campanella came to assert that knowledge of ourselves can only be obtained if it is not obscured by knowledge of the other things that press upon us. The way the soul originally understands itself is an 'intelligere abditum', while the knowledge we have of things other than ourselves is 'addita' or 'illata'. The latter could prevent us from fully knowing ourselves if we were allowed to bury the 'intelligere abditum' under layers of information coming from outside.

Consciousness indicates self-consciousness and 'the perceiving of perceiving' (Leibniz's 'apperception'), that is, the awareness of living and seeing oneself living. But it also indicates the whole world that moves within man and animates him in rational and irrational directions, determining the differences between human beings, each of whom lives unique and unrepeatable subjective experiences precisely because of this intimate and irreducible core that contains probably the most mysterious and elusive part of the universe that eludes the rules of physics and logic and is the incessant production of immaterial states in perpetual change.

Between the end of the nineteenth century and the first half of the twentieth century, a part of philosophy took an 'anti-mentalistic' and 'anti-dualistic' stance with marked materialistic and naturalistic connotations, mainly due to the success of English methodological behaviourism, according to which the study of the behaviour of subjects must be based only on empirically observable external data, to the exclusion of any introspective investigation of internal factors that elude perception.

However, cognitive science in the decade of the 1970s reverted to dualistic positions (brain/mind), stating that the science of the mind should also appprove internal cognitive processes through the use of computational models of cognitive activities. According to computational functionalism, advocated by Putnam [2], mental processes should be assimilated with computational processes, with the consequence that, as we have already seen, the mind should be compared to software operating in the brain, which should be regarded as hardware.

At the same time and in parallel, however, materialistic positions of a monistic nature (based on the identity between mind and brain) have re-emerged, which have been upheld by the Australian school of philosophy of mind (represented by Smart [3], Place, Armstrong [4]) and by so-called 'eliminativism' according to which there are no mental entities (Paul and Patricia Churchland) [5]. The same has been affirmed by Daniel Dennet's instrumentalism, according to which the mentalistic dimension is only a tool for analysing and interpreting the human being also from a relational point of view, but it does not assume the dignity of an autonomous reality.

Recently, however, new forms of thought have developed that are again critical of materialistic (also known as reductionistic) theses. It is thinkers such as Nagel [6] and John Searl who have emphasised that states of consciousness characterise the singularity of every human being, which cannot be explained solely by examining external, physical data. They support the conception of consciousness as a unity that encompasses every fact and every object and therefore cannot be degraded to a fact among facts and an object among objects, as consciousness transcends facts and objects.

These thinkers, among whom Chalmers [7] can certainly be mentioned, re-propose the *hard problem of consciousness*, i.e. the mystery of that *quid* that makes every single human being different from the others and places him on a privileged plane of observation of the becoming reality that flows in front of his states of consciousness, which give life to an inner world that develops like a film sequence. The existence of *consciousness is* based on the distinction between *mind* (*res cogitans*) and *brain* (*res extensa*) and is formed through a circular motion reminiscent of Fichte's philosophy, according to which the 'I' is a circle that starts from the 'preconscious I' (in which everything is undifferentiated) and after encountering (*recte* 'bumping into') the thing-in-itself or the 'Non-ego' (in the cognitive process of the surrounding nature, i.e. the *res extensa*), only returns to itself by assuming the form of the 'conscious Ego'. Thus the 'I' becomes self-conscious only when it reunites with itself after overcoming the collision against the limit of the 'Non-Ego'. The collision in fact determines the synthesis between 'I' and 'Non-ego' that results in the formation of consciousness (understood in the psychological sense). However, such a synthesis is only momentary because a perfect coincidence between *res cogitans* and *res extensa* will never be possible, as the latter is unbridgeable and every synthesis achieved is followed by the birth of a new contradiction between 'I' and 'Non-Ego' and so on ad infinitum. For no matter how far the effort of the 'I' is pushed towards the task of enclosing within itself the synthesis of the whole, there will always be something outside itself that limits it.

In the second half of the twentieth century, scientific studies on consciousness intensified, thanks mainly to a conference in 1994 in Tucson (Arizona) on 'Science and Consciousness' attended by many scientists and philosophers.

The fact that in the past the scientific world did not attach importance to consciousness is due, according to a thesis put forward in the book 'Galileo's Error' by Goff [8], to Galileo's subdivision of phenomena into two classes: primary qualities (size, shape, space–time location), which are objectively and mathematically measurable, and secondary qualities (colour, taste, etc.), which are subjective. The latter were not studied by science for many years precisely because they were considered secondary by Galileo and could not be grasped through mathematical science, while physics concentrated on the primary qualities because they were objectively calculable. Hence, according to Goff [8], the problem of consciousness, which concerns precisely those subjective qualities on which individual experience determines the singularity of each human being, was neglected for very many years.

According to Goff [8], it is difficult to describe consciousness based on external data. Even Artificial Intelligence is programmed as if it were a self-conscious subject capable of subjective experiences, but in reality it is not, so it might be difficult to distinguish a human being from a robot by external behaviour alone. One cannot know the inner, subjective life of others because it is impossible to see others from their inner world (inner life), which can be full of life (as in humans) or totally dark (as in robots).

To explain what consciousness is, Goff [8] recalls the example known in philosophical circles as 'Black and White Mary': Mary has always been confined to a monochrome room and therefore has never seen colour. All she sees are black and white and grey shadows. Mary is persuaded to study everything written about colours and becomes able to authoritatively describe, for example, the yellow colour of a banana without ever having seen it herself. However, when Mary leaves the room and sees colours in the world of experience she knows something about the colour of the banana that no book had been able to describe. This is the result of her subjective experience (conscious experience), i.e. a quid pluris compared to what a book can represent. No book can describe that subjective experience because it derives from the subject's inner life, which cannot be known by others. The same concept is explained in Nagel's famous study "What is it like to be a bat?".

Daniel Dennet argues that consciousness is only an illusion produced by the brain, but other philosophers (such as Chalmers) believe that experiencing an illusion is also a subjective experience that falls under the concept of consciousness.

The concept of consciousness assumes that reality (as argued by Descartes) is composed of material things (among which are the brain and the human body) and immaterial things (minds and consciousness): the greatest mystery is that of the interaction between brain and consciousness, i.e. between the material world and the immaterial world of consciousness.

The proof of this dualism could come today from quantum physics because even in the reality hitherto known to physics, a duality arises: in the quantum world, the laws of physics do not apply because everything in that world is subject to probability. The phenomenon of 'quantum entanglement', which occurs in the sub-atomic world, could be indicative of the fact that reality is not only composed of matter: two entangled particles are able to exchange information with each other even light years apart, as if they were both communicating via non-material channels and had consciousness. So quantum physics should lead us to think that reality is not just made of matter, because the intrinsic nature of things would be spiritual, as advocated by panpsychism, according to which consciousness pervades everything and is the essence of all reality. In the journey into the infinitely small, the most important step was the penetration into the world of atoms: "by exploring the interior of the atom and studying its structure, science went beyond the limits of our sensory imagination. From this point on, it could no longer rely with absolute certainty on logic and common sense. Atomic physics enabled scientists to take their first quick glimpse into the essential nature of things. Like the mystics, physicists were now dealing with a non-sensory experience of reality and, like them, had to deal with the paradoxical aspects of this experience. From then on, therefore, the models and images of modern physics became similar to those of Eastern philosophy'.[11] Moreover, the further we move away from sensory experience to know reality, the more inadequate our ordinary language becomes to describe what we encounter when penetrating deeply into the nature of things.

Instead, the materialist conception focuses on how things interact with each other and how bodies behave according to the laws of physics, but fails to say what the essence of things is and what they intrinsically are. It merely describes matter and explains phenomena in deterministic and mechanistic terms on the basis of relations

[11] The Tao of Physics, Fritjof Capra, Gli Adelphi, 1982.

of forces and energy in motion and causal relations. Thus science only describes the extrinsic properties of matter, but says nothing about its intrinsic characteristics (inner nature). Goff [8] uses the example of the chess piece: knowing what a bishop is doing on a chessboard (moving diagonally) does not imply knowledge of the matter of which that piece is made. Magritte's famous painting 'The Treachery of Images' ('ceci n'est pas une pipe') also expresses the same concept.[12]

The problem of the relationship between brain and mind has been addressed not only in the philosophical field but also in the scientific field (with particular reference to biotechnology and neuroscience) and psychiatry. The positions in the latter field that are divided on the concept of mental illness are well known: according to the materialistic or biological position (so-called somatic psychiatry, whose first exponent was Philippe Pinel in the first half of the nineteenth century), it is not possible to conceptually separate the mind from the brain, with the consequence that many of the operations that we define as mental are nothing more than the result of the physiological functioning of the brain. Therefore, it is not appropriate to speak of mental illness because it is the damage to the brain and nervous system that produces those dysfunctions that some call mental problems. For example, damage to the brain can result in loss of memory and self-awareness or a change in an individual's identity and personality, or the alteration of so-called moods and related behaviour[13] [9].

Somatic psychiatry (whose major proponents include Antoine L. Bayle, B.A. Morel, W. Griesinger) considers the brain as the organ of all soul functions. This scientific approach is also called biological psychiatry and, according to the philosopher Dominic Murphy, it consists of psychiatry being considered a branch of medicine dedicated to discovering the neurological basis of all so-called mental dysfunctions. The philosopher Churchland [5] has also pointed out that problems relating to the psychic sphere, often referred to as mental disorders, must be traced back to the physical sphere in a unified scientific framework based on the physiology, chemistry and genetics of the brain. The preference for the materialistic view marks a departure from Freudian psychoanalysis, which is now regarded by many as a non-scientific discipline[14] [9].

Somatic psychiatry focused mainly on cases in which the psychic disorder was believed to be a consequence of an organic injury, but extended this causal relationship to hypotheses in which organic conditioning was not empirically observable. Many psychiatrists realised the lack of scientific foundation for this belief, as many forms of neurosis (hysteria, paranoia, mania, hypochondria) that had no connection with an organic lesion were examined, with the consequence that the aetiological connection between psychic and somatic fact could not always be traced back to universal and necessary laws. This conclusion was also confirmed by the study of hypnosis, which

[12] Panpsychism and the problem of consciousness, by Richard Gault, Beshara Magazine, issue 15/2020.

[13] Mind—Brain Dualism and its place in mental health care, issued in Handbook of the philosophy of medicine, 21 May 2015, by Eric Matthews.

[14] Mind—Brain Dualism and its place in mental health care, issued in Handbook of the philosophy of medicine, 21 May 2015, by Eric Matthews.

can provoke suggestions whose symptoms can be assimilated to those generally resulting from organic lesions.

With E. Kraepelin, between the end of the nineteenth century and the beginning of the twentieth century, psychiatry followed a different path with respect to the somatic conception, arguing that the mind/brain dualism is fundamental to understanding mental phenomena, since, while the brain reacts to external stimuli through the functioning of the neuronal apparatus in an identical manner in all individuals, thoughts, emotions, desires and other mental phenomena depend on properties that the brain cannot possess: intentionality and subjectivity. They are in fact always subjective and singular states that are based on an immaterial *quid that* cannot be replicated either in different individuals or in a scientific experiment (unlike many brain functions that are reproducible in a laboratory). Furthermore, intentionality indicates that the mind (unlike the brain) expresses thoughts and emotions in relation to something external to it, with which a non-physical relationship is created that revolves around the individual's past experiences and inner psychic state (it is precisely the collection of these subjective experiences that constitutes the essential core of consciousness). The mind/brain dualism asserted by these psychiatrists induces them to solve mental disorders that are not dependent on organic problems by resorting to non-drug therapies. Furthermore, they consider mental dysfunctions as a set of behaviours that are not empathetic to the surrounding environment and that depend on an individual's existential malaise, the cause of which may derive from the social reality that is hostile to him. The psyche is considered to be the function by which the living organism enters into relationship with the environment, thus an individual's mental disorder is often determined by harmful social relationships in relation to which subjective behavioural responses are (improperly) classified as pathological, with the consequence that the treatment of the patient must take place by taking into account his relationships with the outside world rather than by searching for an internal neurological or cerebral dysfunction. This school of psychiatric thought reverses the cause/effect relationships that have dominated materialistic culture: if a person has behavioural problems that affect his family and social relationships, the cause is not always a neurological or otherwise organic malfunction, but could instead be an attempt to defend himself against abusive external pressures.[15]

With E. Bleuer, between the end of the 19th and the beginning of the twentieth century, psychiatry began to be sensitive to the influence of depth psychology (a concept later developed by Jung). "Schizophrenia is considered to be a splitting of the mind consisting of a dissociation of psychic functions (whereas previously it was considered to be a degeneration of the intellect with an organic basis). Connected to schizophrenia is autism, i.e. isolation of the mental-subjective dimension and separation from the real world with shattering of the fantasy world into a multiplicity of concepts separated from one another".[16]

The further development of psychiatry moves towards overcoming dualism between physical and psychic facts. This happens in the twentieth century with Karl

---

[15] Emanule Severino, La filosofia dai greci al nostro tempo—La filosofia contemporanea BUR 1996.

[16] Emanule Severino, La filosofia dai greci al nostro tempo—La filosofia contemporanea BUR 1996.

Jaspers, who in his 'General Psychopathology' argues that 'man is an inseparable totality that must be brought to light not by the methods of medical science but by those of phenomenoloy'.[17] In Ludwig Binswanger, the overcoming of all dualism goes so far as to marginalise the very opposition between health and illness, normality and abnormality. Phenomenology, especially in the form it assumed in Heidegger, shows, beneath that opposition, the existence of man whose madness is not to be understood as a pathological deviation from normality, i.e. as a disorder that alters mental health and which must be explained and treated according to the criteria of clinical psychiatry, but with another kind of order, in which a mode of existence is manifested and realised, i.e. one of the real possibilities in which human existence consists'.[18]

In the field of science, studies have mainly focused on the functioning of the brain. However, despite the vast amount of information acquired on this organ, understanding its complex functions is still a mystery because each new discovery opens up multiple avenues from which new problems to be solved and complicated mechanisms emerge with increasing levels of difficulty. So much so that, according to some scientists, the brain cannot be defined as a single thing or a single organ but rather as a set of interconnected patterns in a unitary functional framework.

The metaphor of the computer to define the brain, which, as we have seen, has often been used to explain its functions on the basis of the paradigms that govern computer processes, may turn out not to be satisfactory, since, in addition to the divergence between the natural element and the underlying artefact, it has been pointed out that the brain is the result of millennia of evolution developed on experience and interaction with other living beings, whereas any kind of artificial intelligence cannot possess that same depth and complexity. The computer is like a flat mirror: it receives input, passively absorbs stimuli and reacts by representing and processing information. Whereas man not only reacts to external stimuli but also acts by constructing information and exercising an autopoietic function through the generation of abstract and universal concepts thanks to thousands of years of evolutionary history, pathos, logos, psychic entanglements and irrational components, thus giving rise to some of his most mysterious and beautiful characteristics: singularity and self-awareness (which implies the awareness of living and dying). Compared to the invariability of computers, which all produce the same reactions to input, man (Nietzsche's too-human human) is distinguished by the multiplicity of his behaviour and thoughts. If we placed 10 objects with artificial intelligence on a beach in front of a ship crossing the sea at sunset, their 'sensations' would all be identical. The same image seen by 10 human beings would unleash not only 10 different kinds of sensations but a myriad of multifarious sensations depending on the biological, historical and cultural background of each individual, from which would arise other new sensations and thoughts that never existed before. And if among those 10 human beings were a poet, who knows how many secrets hidden behind that ship and landscape would be revealed.

---

[17] Emanule Severino, La filosofia dai greci al nostro tempo—La filosofia contemporanea BUR 1996.

[18] Emanule Severino, La filosofia dai greci al nostro tempo—La filosofia contemporanea BUR 1996.

1.4 Transhumanism and Posthumanism

Transhumanism is an intellectual movement based on the use of technology to radically transform the human organism in order to enhance its intellectual and physical faculties with the aim of overcoming human limitations and moving it towards a posthuman era in which human beings will have such enhanced cognitive, emotional and physical capacities that they can no longer be considered human.

The proponents of this movement are inspired by humanistic and scientific cultural traditions, with particular reference to ancient philosophy and Enlightenment rationalism, as they envisage a gradual transition from human to posthuman and not an evolutionary discontinuity.

Although the intentions of transhumanists would seem to be to improve human living conditions, one should ask oneself whether it is really necessary to artificially influence the natural evolutionary process of mankind, proceeding in the dark along a road that could present bitter surprises and dramatic outcomes. On the other hand, if the natural development of life has brought mankind this far, would it not be more prudent to continue letting things flow spontaneously without forcing things? If the subject has always been man and the predicate every object of his creation (including artificial intelligence and technologies), why risk reversing these relationships?

It was precisely these questions that gave birth to bioethics, whose name originated in 1971 from the title of a book by the carcinologist V.R. Potter, who 'intended to build a bridge between the life sciences and ethics'[19] to prevent biotechnology and medical and pharmaceutical techniques from altering the most significant passages of life (conception, birth, paternity and maternity, illness, suffering, death) without serious ethical reflections underpinning them, taking into account the implications of affective and sentimental naura and the legal implications relating to human rights, the person, identity, dignity and social relations.

In fact, scientific progress has already initiated a transition towards something that is a quid pluris with respect to man. One thinks of procreation in the laboratory by means of artificial fertilisation techniques, thanks to which what was dictated by necessity is now the subject of choice, with the consequent possibility of generating children even at a late age or in cases where this would naturally be impossible. One thinks of the manipulation of stem cells, which could make it possible to regenerate diseased tissue or could allow the artificial creation of organs to be implanted in the human body.

Should the path to posthumanism be deemed irreversible, bioethics could act as a guide in this transition to at least protect and pass on the values derived from *humanitas*.

---

[19] The Philosophy of the 20th Century, Remo Bodei.

1.5 An Epilogue

Based on what has been said so far, our intelligent thing may not have found any answers to the questions about its origin and the purpose of its existence.

However, if it were designed to conceive paradoxes,[20] it would come close to some truth. Our intelligent object could convince itself that its extremely rational being was born from its opposite, namely the most irrational part of man and the darkest and most unfathomable sides of the human mind.

After all, the world is one big irrational[21] and revolves around a series of contradictions that form its foundation. Just think of opposites and their interchangeability.

Perhaps the birth of artificial intelligence is due to man's need to still take refuge within the walls of logic and reason when, in the course of his history, he realised the chilling image of a universe without purpose and indifferent to his pathos [10]. Hence he still felt the need to anaesthetise his torments by making them flow into logic rather than immerse himself in becoming (as Nietzsche had suggested he should do) and traverse the infinite field of the irrational to the far reaches of life and death. Perhaps, like Sisyphus, man has seen himself imprisoned in a labyrinth of mirrors in which everything appears incomprehensible, and has felt the urge towards something dominatable with logic and reason: 'I want everything explained to me or nothing. And reason is powerless before this cry of the heart. The spirit, awakened by this demand, seeks and finds only contradictions and unreason. What I do not understand is without reason. The world is peopled by these irrationals and the world itself, whose unique meaning I do not understand, is nothing but an immense irrational…At this point in his endeavour, man is confronted by the irrational and feels within himself a desire for happiness and reason. The absurd arises from the confrontation between the human call and the unreasoning silence of the world. It is this that one must not forget; it is this that one must cling to, for the consequences of a whole life can be born. The irrational, human longing and the absurd, which arises from their intimate conversation: these are the three characters in the drama that must necessarily end with all the logic of which an existence is capable'.[22]

The creation of artificial intelligence could therefore be the epilogue to this drama experienced by man on the tightrope above the abyss: instead of walking it all the way and saying an unconditional yes to the metamorphoses of life and to the multitudes that inhabit his psyche, overcoming the enclosure of reason and illuminating the secret and rapacious places of the universe where only the poetry and madness of genius are able to reach, man seems to have preferred to put on the straitjacket of logic once again, generating a miniature of himself, devoid of chiaroscuro, of nuances, of emotional contractions and dilations, of artistic and creative flashes, of imaginative power that produces synaesthesia, of psychic figures that populate the unconscious and nourish life with multiform and new motives and multiple themes of existence.

[20] Paradox comes from para—doxa, i.e. against common opinion.

[21] Albert Camus, The Myth of Sisyphus.

[22] Albert Camus, The Myth of Sisyphus.

In the great tree of human thought a monstrous fruit has ripened and it represents the reduction of the tree itself.[23]

References

1. Mc Carthy, J.: AA VV, Semantic Information Processing. Mit Press (1968)
2. Putnam, H.: Representation and Reality (1991)
3. Smart, J.: Our Place in the Universe. A Metaphysical Discussion (1989)
4. Armstrong, D.: The Nature of Mind (1980)
5. Churchland, P.: Matter and Consciousness: A Contemporary Introduction to the Philosophy of Mind (1988)
6. Nagel, T.: Mind and Cosmos (2012)
7. Chalmers, D.: The Conscious Mind (1997)
8. Goff, P.: Galileo's Error (2020)
9. Matthews, E.: Mind—brain dualism and its place in mental health care. Handb. Philos. Med. (2015)
10. Blakemore, C.: AA VV, Human Mind and Artificial Mind. Feltrinelli, Milano (1989)

[23] Ecce Homo F. Nietzsche—Fatal Monologue by Robeto Galasso (Adelphi 1991).

Chapter 2
A Puppet Without Strings? Two Considerations on AI and Its Ethics

Domenico Marino, Daniele Cananzi, and Filippo Aragona

2.1 AI at the Test of Thinkability

One of the best-known stories, and not only in children's literature, is that of the adventures of Pinocchio, the character Carlo Collodi gives birth to in an absolutely analogical world in which the ancestral desire to create an automaton with human features is realised in the form of a wooden puppet—not human, therefore—but very much affected by the faults, all human and all too human, which can perhaps be summed up in 'doing the wrong thing'.

There has been constant reference to Pinocchio in the context of studies on artificial intelligence (a sort of Pinocchio 3.0) and many have picked up on a similarity between that wireless puppet and certain robotic experiments that attempt to build a machine not only without hardware connections (the wires) but above all without software connections (the programming): artificial intelligence is perceived as a 'thinking machine', to use one of the widely used expressions, in the sense that it should be, one day at least, capable of going beyond its own programming, both in the sense of learning to learn (and on this the technology has advanced), and in the sense of being able to make decisions in a further and original way with respect to the data and functions initially provided.

To such an extent that the Collodian idea—a highly original evolution of that long series of representations that have dotted human history (the myth of the Golden Handmaids of Ephesus, Pygmalion, Vaucanson's mechanical duck, Droz's automata, etc.)—seems to have made this definitive evolutionary leap (as in the case of AlfaZero for the game Go[1] [1]), taking concrete form in systems capable of exceeding programming and, it would seem, of thinking autonomously, learning from experience, arriving at predictive decisions on reality, as in the cases of AI

[1] Cf. Aa. Vv., *A general reinforcement learning algorithm that masters chess shogi and Go through self-play*, in Science, 2018, no. 362, p. 1140 ff.

D. Marino et al., *Ethics and Artificial Intelligence*, Studies in Systems, Decision and Control 519, https://doi.org/10.1007/978-3-031-50902-5_2

applied to judicial activity[2] [56], or responsible actions, as in the case of driverless cars[3] [33] or autonomous weapons[4] [3].

Between desire and reality, between epochal technological revolutions and fanciful reconstructions still too closely tied to myth and imagination (utopian or despotic), AI presents itself as a theme from which it is difficult to disregard also because, effectively, we are surrounded and, in some cases, controlled by it, on a daily basis.

In what terms is artificial intelligence thinkable today? And what kind of ethics are we talking about when we talk about the ethics of artificial intelligence? The following pages focus on this twofold question and, briefly, propose to develop two simple considerations in the light of the current international debate—now much absorbed precisely by the subject of artificial intelligence—which could be summarised in the grid of positionings that Max Tegmark[5] has hypothesised in his *Life 3.0 in which he* proposes, on the one hand, the 'digital utopians', those who are prepared to think of an authentic Copernican revolution with AI reaching the extremes of complete autonomy and freedom, and, on the other hand, the 'techno-skeptics', those who do not believe it is possible that the myth of the creation of a free non-human being that is totally autonomous will ever come to pass.

The following pages do not take sides in either direction because they discuss an assumption, to which they attempt to give some argument, believing that this opposition between sceptics and supporters is not, perhaps, the way to take up the issue, and they deal with what AI allows us to say today. The essay, therefore, reflects on the preliminaries of possible reasoning about AI and tries to identify the contribution that AI technology can already make today to the debate, not only to that devoted to new technologies but also, and perhaps above all, to the political, legal, medical and ethical ones.

The analysis proceeds—we should immediately warn—with the necessary simplifications that, we hope, will not invalidate the discourse in its theoretical framework. These simplifications are necessary in order to provide a homogeneous thread of argumentation on the various fields of research in AI which, at present, appear much more complex, different and diversified, as much in content and in the individual fields as in the different technological progress that distinguishes them. It should be noted, therefore, that the realities that are the subject of interest and the dimensions in which AI is developing appear to be much broader than the ones outlined here, and contain a myriad of specificities that make them at times scarcely compatible with a broader and more general (though hopefully not generic) discourse such as the one we intend to propose.

---

[2] Cf. G. Sartor, *L'intelligenza artificiale e il diritto*, Turin, 2022, p. 122 ff.

[3] See P. Lin, R. Jenkins, K. Abney, *Robot Etics 2.0. From Autonomous Cars to Artificial Intelligence*, New York, 2017.

[4] R.C. Arkin, *Governing Lethal Behaviour in Autonomous Robots*, Boca, 2009.

[5] M. Tegmark, *Life 3.0*, Chapter 1.

2.2 Artificial Understanding of Reality: How Intelligent is AI?

While it is true that one of the characteristics of AI studies is that they contain many different and often opposing directions, it is equally true that not even on the qualification of artificial intelligence has there been general agreement[6] [32].

Of those proposed, it seems to me that the most appreciable is—and continues to be—that of McCathy *et alii* in 1955 [36]: 'The problem with artificial intelligence is to make a machine act in ways that would be called intelligent if a human being behaved in the same way'.[7]

Unlike so many other definitions, this one clarifies the terms of what it means to hypothesise a non-human intelligence and perhaps also allows us to appreciate not only the future dystopian or utopian (depending on one's point of view) perspectives that so many are in a hurry to elaborate, but to consider how much we are already immersed (even unconsciously) in the new reality[8] [31]. In fact, it is not a question of reworking Orwell-style dystopias with a technological component, but of taking into account established, or at least widespread, realities, such as autonomous cars, autonomous weapons, the robotic surgeon, machines for the permanent assessment of health and the administration of pharmacological therapies, stock exchange algorithms that buy and sell securities on the markets, etc.

The promises of future applications are nothing more than evolutions of our current reality. For example? A world in which the quality of life is greatly enhanced by AI: new forms of workforce, new forms of assistance to everyday life, new forms of relationality; just think of AI applied through the algorithms of love that bring people together without all that traditional rigmarole of enchantment-courtship-love that has marked pages of literature and filled libraries and cinemas.

It is in this direction that Tehmark goes, in the *Life 3.0* study already mentioned, in which he envisages a Life 3.0 that would follow Life 1.0 and Life 2.0 evolutionarily. Life 1.0 would be that *biological stage of* evolution in which, as he writes, hardware and software are subject only to biological evolution; Life 2.0 would be that *cultural stage* in which hardware is subject to biological evolution while software is designed culturally; Life 3.0 would be that *technological stage* in which both hardware and software are designed technologically.

If intelligence can be defined as the 'capacity to achieve complex ends', then it is well understood that AI is, and will be, a non-biological and non-human intelligence.

Let us give three examples:

(1) the logarithms of love; (2) legal disputes; (3) automatic weapons.

---

[6] Fifty-three different definitions are given in S. Legg, M. Hutter, *A collection of intelligence*, in B. Goertzel, P. Wang, eds., *Advances in Artificial General Intelligence: Concepts, Architecture and Algorithms*, Amsterdam, 2007, p. 17 ff.

[7] J. McCathey, M.L. Minsky, N. Rochester, C.E. Shannon, *A proposal for the Darthmouth summer research project on artificial intelligence, august 31, 1955*, in 'AI Magazine', 27, 2006.

[8] Cf. F. Lazzini, *Digital Ethics and Artificial Intelligence,* Turin, 2023, p. 23 ff.

In each of these examples, artificial intelligence works by computing big data, i.e. by linking data with statistical logic and in a statistical manner.

As Gigerezen[9] [20] reminds us, the logarithms employed by dating sites and agencies process the data deduced from the profiling of subjects and trace their correspondences through *similarities*, *complementarities* and *importance*; the advertising slogan of one of the most representative is effective and indicative: '*Every 11 min, a single person falls in love*'.

These algorithms, so incredible, at least in their claimed promise of finding love, and so ambitious, in their basic premise of explaining what love is, are not disclosed; but their workings are clear and consist of having people answer a profiling questionnaire and then matching individuals with a high percentage of matches based on the detected compatibility that results from the numbers. The system, in fact, turns the data entered by the individuals into numbers and processes these numbers statistically. The promise of love, therefore, is translated into the *satisfaction that is* obtained based on the profiling. Of course, it would seem a neutral way of analysing two people and attempting to objectively track compatibility. In reality, however, neither the way is neutral, nor does the profile match the person, quite the contrary. On the first aspect, it suffices to note that the compatibility between people and the satisfactory and lasting success of their meeting that turns into a relationship does not depend only on certain preferences and tastes, it depends on information that affects them deeply and leaves ample room for the irrational and instinctive, just think of looks, scents, tone of voice, etc.; the alleged neutrality, then, is nothing more than the putting in brackets of these elements, perhaps not secondary after all. The fact that the profiling does not correspond to the person is a second noteworthy element; not only because in answering the questionnaire there is a high percentage of intentionally falsified answers (height, weight, personality traits, etc., which—usually—are sweetened to represent oneself better than one is), but, above all, the person is not the sum of his or her characteristics while the profile is precisely the sum of his or her information[10] [16].

Not dissimilar considerations can be made in the description of AI and logarithms applied to another human dimension capable of warming minds, legal conflict, the second area to focus on. Tegmark asks what mental associations come to mind when thinking about the administration of justice and answers with long delays, high costs and occasional wrongful convictions; he adds: "wouldn't it be wonderful if the first thoughts were 'efficiency' and 'fairness'?".[11]

Here too, the application of AI could have a positive effect: on fairness because the decision-making system would have no emotional, environmental, relational or ethical influence; on efficiency because a decision could be reached quickly

---

[9] G. Gigerenzer, *How to Stay Smart in a Smart World. Wy Human Intelligence Still Beats Algorithms*, London, 2022.

[10] See E.J. Finkler, P.W. Eastwick, et alii, *Online dating. A critical analysis from the perspective of psychological science,* in Psychological Science in the Public Interest, 2012, no. 13, p. 3 ff.

[11] M. Tegmark, *Life 3.0,* cit., Chapter 3.

without the lengthy time taken by judgement but by employing the 'competence' and 'transparency' of AI.

As Tegmark also notes, 'These are important questions that we all need to ponder and discuss to ensure that AI remains beneficial. We aren't facing an all-or-nothing decision regarding robojudges, but rather a decision about the extent and speed with which we want to deploy AI in our legal system. Do we want human judges to have AI-based decision support systems, just like tomorrow's medical doctors? Do we want to go further and have robojudge decisions that can be appealed to human judges, or do we want to go all the way and give even the final say to machines, even for death penalties?".[12]

While waiting to decide on the future, we can say two things about two realities of the present. The first is *Claudette*, a machine learning system that identifies and classifies potentially unlawful or abusive clauses within contracts or privacy notices[13]; although there is still much to be improved, as stated in a study on the privacy policies of large economic groups with social impact,[14] "our study suggests that the fields of legal informatics and applied machine learning can be significantly pushed forward"[15] [9, 55]. The functioning of Claudette is well described by Giovanni Sartor: "the system, based on training, generates its own concept of vexatiousness, which it then applies to the classification of new clauses. This concept is constructed in such a way that, in applying it, the system reaches the same conclusions as those reached by the jurists who prepared the training set, with respect to the cases contained in that set".[16] Here too, one must first transform the text of the clauses into data that can be functional for the analysis; here too, as in the preceding case of the logarithms of love, it is not so much the clause in and of itself that is being examined as its *datafication*, which is independent of the semantic structure of the sentence.[17]

The second current reality in litigation, the one to which Tegmark's remarks also referred, is that of *predictive justice*. Of course, even in this case, one has to wait for the constant evolution of the available technology, but already now one can see how systems such as COMPAS[18] [15, 40] are being used for the assessment of bread or for the recognition or not of liberty in the case of recidivism. Suffice it

[12] *Ibid*.

[13] A prototype of the system can be used at http://claudette.eui.eu/.

[14] In particular: Google, Facebook (and Instagram), Amazon, Apple, Microsoft, WhatsApp, Twitter, Uber, AirBnB, Booking.com, Skyscanner, Netflix, Steam and Epic Games.

[15] G. Contissa et alii, *Claudette Meets GDPR: Automating the Evaluation of Privacy Policies Using Artificial Intelligence,* in "SSRN- Electronic Journal", January 2018, p. 58.; see also F. Ruggeri et alii, *Detecting and explaining unfairness in consumer contracts through memory networks*, in "Artificial Intelligence and Law", 2022, no. 30, p. 59 ff.

[16] G. Sartor, *L'intelligenza artificiale e il diritto*, cit., p. 125.

[17] As it does not affect the syntactic structure of the sentence; to take the example proposed by Sartor: "the sentence 'the user is always responsible' is represented by the vector [00100000111]", ibid. p. 126.

[18] The issue is widespread and other similar systems are Hart or OxRec on which, respectively: M. Oswald *et alii*, *Algorithmic Risk Assessment Politicing Models. Lessons from the Duhram HART Model and 'Experimental' Proportionality*, in "Information and Communications TechnologY Law", 2018, no. 27, p. 223 ff.; S. Fazel et alii, *Preciction of Violent Reoffending in Prisoners*

to recall the case *State of Wisconsin v. Eric L. Loomis*[19] in which the US Supreme Court recognised the admissibility of the use of COMPAS, albeit with limitations and caution in the use of the software[20] [29]. Here, too, the management of the mass of data available makes the system capable of expressing a decision that does not focus on the specific case but is established in the light of the totality of the case-precedents examined. The considerations of two judges in particular, who adhered to the majority opinion, are interesting. Judge Drake Roggensack observes: "I agree with the majority opinion that circuit courts may consider a COMPAS risk assessment along with a multitude of other relevant factors at sentencing, as was done in this case. (…) The majority opinion concludes that '*if used properly with an awareness of the limitations and cautions*, a circuit court's consideration of a COMPAS risk assessment at sentencing does not violate a defendant's right to due process. I agree that "consideration" of COMPAS does not contravene defendant's right to due process. (…) Stated otherwise, rather than merely considering COMPAS as one of many factors relevant to sentencing, the majority opinion, due to its interchangeable *use of "rely" and "consider,"* together with the certified question, may be read to permit a circuit court to rely on COMPAS to determine the appropriate sentence. Reliance *would violate due process protections*. Accordingly, I write to clarify our holding in the majority opinion: *consideration of COMPAS is permissible; reliance on COMPAS for the sentence imposed is not* permissible".[21] Justice Shirley S. Abrahamson points out: "I join the majority opinion. It describes the salient issues raised by considering COMPAS at sentencing, and informs the bench and the bar of the limitations and cautions that should be observed in considering COMPAS in sentencing. It underscores that *we are addressing the use of a research-based tool* and that it is incumbent upon actors in the criminal justice system to recognise that additional research data may become available in the future and different, better tools may be developed. (…) I conclude that in considering COMPAS (or other risk assessment tools) in sentencing, a circuit court must set *forth on the record a meaningful process of reasoning addressing the relevance, strengths, and weaknesses of the risk assessment* tool".[22]

The combination of the two opinions shows how the Supreme Court, in confirming the legitimacy of the use of COMPAS, recognises it as an assessment *tool* and not as a method of assessment; a tool to be used with its own *cautions* and *limits*, in the essential distinction between 'considering' the tool, a possible and legitimate operation, and 'relying' on the tool, an illegitimate operation; use of the tool that must be justified in the justification by illustrating its strengths and weaknesses.

---

and Individuals on Probation. A Durtch Validation Study (OxRec), in Scientific Reports, 2019, no. 9, 841, p. 1 ff.

[19] Case No. 2015AP157-CR, 5 April-13 July 2016; the judgment can be found at https://law.justia.com/cases/wisconsin/supreme-court/2016/2015ap000157-cr.html.

[20] Cf. F. Lagiola, R. Rovatti, G. Sartor, *Algorithmic fairness through group parities? The case of COMPAS-SAPMOC*, in AI and Society, 2023, 38, p. 462 ff.

[21] *State of Wisconsin v. Eric L. Loomis*, pp. 3–4, italics mine.

[22] *State of Wisconsin v. Eric L. Loomis*, p. 1, italics mine.

A classic distinction that would seem to be confirmed, that between judgement and decision, of primary importance in the reconstruction of procedural structures and that, while enriching the jurist's tools with a new means, requires a considered justification of the same tool in the reasoning of the justification[23] [37].

On the other hand, one concern that cannot be overlooked is the suspicion about the assumed neutrality of the so-called robojudge; acting on the processing of data from previous cases, it remains stuck on possible biases that have influenced said precedents as evidenced by those who emphasise the prejudice against African-Americans in a series of judgments that are therefore completely unfair[24] [38].

To this must be added that, in any case, as things stand, the predictive system always acts in a 'blind' manner,[25] i.e. without any awareness or knowledge of the facts of the specific case. Which means, like love algorithms, assuming an evaluation that affects the case but not the case, but the computational combinatorics on previous cases and stored data, which affects the present but not the present, rather the past of previous cases and stored data.

The essential judgement/decision distinction and the considerations proposed last are valid and remain fundamental even in the third example-area that we will briefly consider: autonomous weapons.

Even on this issue, there is still no generalised agreement on what is to be understood as an autonomous weapon; They range from broader qualifications, such as the 2012 definition of the US Department of Defence, for which the autonomous weapon is one that is capable of selecting and attacking the target without human intervention,[26] to the more restrictive 2017 definition of the British Ministry of Defence, for which the autonomous weapon must include high-level intentions and directives in order to take actions to achieve the target,[27] to the more normative definition of the Swiss government, for which the autonomous weapon must perform tasks that are respectful of international humanitarian law.[28] Guglielmo Tamburrini, in an attempt to clarify the issue and identify the real focus of the problem, points out that from

---

[23] That this instrument then goes beyond the legal-processual component alone is soon observed with the gender leap to politics as well; as Tegmark again observes, "But should AI-based decision support systems for voters and legislators ensue, followed by outright robolegislators?", M. Tegmark, *Life 3.0*, cit., Chapter 3. See also F. Musiani, *Algorithmic governance: surveillance, censorship and fundamental rights*, in F. Fossa, V. Schiaffonati, G. Tamburrini, eds., *Automi e persone*, Rome, 2021, p. 95 ff.

[24] See M. Tegmark, *Life 3.0*, cit.; C. O'Neil, *Weapons of Math Destruction. How Big Data Increases Inequality and Threatens Democracy*, New York, 2016.

[25] On this point see G. Sartor, *L'intelligenza artificiale e il diritto*, cit., p. 135.

[26] US Department of Defense, *Autonomy in Weapons System*, directive 3000.09, 2017, at www.dtic.mil/whs/directives/corres/pdf)300009.pdf.

[27] UK Ministry of Defence, Unmanned Aircraft System, 2017, at https://assets.publishing.service.gov.uk/government/uploads/system/uploads/attachment_data/file/673940/doctrine_uk_uas_jdp_0_30_2.pdf.

[28] United Nation, Towards a "Compliance-Based" Approach to Laws. Informal Working Papar Sudmitted by Switzerland, 2016, in https://docs-library.unoda.org/Convention_on_Certain_Conventional_Weapons_-_Informal_Meeting_of_Experts_(2016)/2016_LAWS%2BMX_CountryPaper%2BSwitzerland.pdf.

the problem of defining the autonomous weapon, we have more profitably moved on to consider the relationship with human control over the work of the AI: 'The British NGO Article 36 has the merit of having first identified the human element as the key issue to be addressed in the reflection on autonomous weapons. (...) it has emphasised the need to ensure that attacks by all weapon systems are subject to 'meaningful human control'"[29] [61]. This, if on the one hand—as Rebecca Crootof[30] [10] notes—has manifested itself as an 'intuitively appealing principle'[31] of which to even discuss a sort of customary nature, on the other hand, the definitional question remains open as to what is meant by meaningful human control, given that various things can be meant by this formula, to such an extent that—given the difficulty in pinning down what -meaningful human control‖actually requires,—[s]everal states [have] expressed skepticism over the added value of the suggested concept, assessing it as being too vague, subjective and unclear‖.[32] Nonetheless, Crootof believes that, in this case, the vagueness of the formula may be an advantage since the vagueness makes it more likely to be incorporated into international agreements and treaties that are based on consensus; Crootof adds: "perhaps because of-its inherent imprecision, a proscription on weapons incapable of being meaningfully controlled might usefully join existing prohibitions on new weaponry. While incorporating this principle in States' legal reviews will not address all of the issues raised by AWS, it will help concretise what meaningful human control actually means through State practice. However, this principle will only be beneficial if it does not undermine existing humanitarian norms governing targeting".[33]

Admittedly, the concept and idea of 'human control' goes a long way towards reducing the scope of the autonomy of the autonomous weapon and reveals how, fundamentally, the moment one moves from the design of an autonomous reality to its practical implementation, it almost seems as if the concession or recognition of a real autonomy of the weapon has to be greatly reduced. But this confirms what was argued in the previous cases examined, the logarithms of love and robojudge, namely that the AI remains a tool that does not replace the human being and does not assume his real autonomy; a tool at the service of improving, respectively, the possibilities of encounters, case examination and military action that reduces human casualties. Even in the case of autonomous weapons, moreover, the considerations made regarding statistical learning of the action to be performed and the AI's limited intelligence to computation of information not to understanding reality and thus possible awareness in action remain confirmed.

But this raises the second question proposed in these pages: to whom is the so-called AI ethics intended to be directed?

---

[29] G. Tamburrini, *Ethics of Machines,* Rome, 2020, p. 105.

[30] R. Crootof, *A Meaningful Floor for "Meaningful Human Control"*, in *Temple International and Comparative Law Journal*, 2016, no. 30, p. 53 ff.

[31] Ibid, p. 53.

[32] Ibid, p. 58.

[33] Ibid, p. 60.

2.3 Opportunities and Risks in the Application of AI: For Whom is the Ethics of AI Ethics?

Putting ethics and AI together, at least at first glance, means being able to argue that the AI is now so intelligent, and that this gives it such autonomy, that the question of the ethical evaluation of its conduct must be raised, thus making the AI a true 'moral agent'. A veritable 'puppet without strings' that—just like Pinocchio from whom these considerations started—acts and can therefore be said to act either well or badly, ethically well or ethically badly: it is bad, it belongs to the ethically bad, the action that Pinocchio performs when he sells the abecedario from Geppetto paid at the high price of the jacket in order to go and have fun with his playmates; it is good, it belongs to the ethically good, the action that Pinocchio performs when he goes to look for his father Geppetto lost at sea. Collodi, as is evident, thinks of a puppet without strings but with all-human capabilities and fallibility. Has the current technological progress really achieved this? Is the new Pinocchio predictable of good and evil or—in Nietzsche's expression—does he still remain 'beyond good and evil'?

Under the banner of ethical correctness and incorrectness, the scientific debate on the ethics of AI has indeed become very wide-ranging, also in view of the actions that algorithms perform on a daily basis; one need only recall the trading of financial markets to understand in what terms the ethical question may be relevant[34] [18]. Also and above all in the light of the consideration, well identified by Floridi, that "algorithms are not ethically neutral"[35] [17], one must then question oneself as to the actual possibility of referring to AI and algorithms the ethicality of action, verifying whether or not the algorithmic 'wireless puppet' is comparable to Collodi's, with all the consequences of the case.

On the one hand, evidence of non-neutrality is not lacking and is also documented in the literature: of the discrimination of legal systems decision-makers has already been mentioned; of the inefficiency in the allocation of advertisements evidence has been amply provided[36] [12, 60] as well as the reason why, for example in the STEP (Science, Technology, Engineering, Mathematics) sector[37] [30], the algorithm designed on the cost-effectiveness logic of distribution ends up discriminating against young women; not to mention the health sector and the better care intended for white patients than for black patients because the algorithm uses the parameter of cost and

---

[34] On this point see A. Foti, D. Marino, *Blockchain and Charities: A Systemic Opportunity to Create Social Value*, in D. Marino, M. Monaca (eds.) *Economic and Policy Implications of Artificial Intelligence. Studies in Systems, Decision and Control*, Cham, 2022, pp. 145–148.

[35] L. Floridi, *Ethics of Artificial Intelligence*, Milan, 2022, p. 146.

[36] L. Sweeney, Discrimination in online ad delivery. Google ads, black names and white names, racial discrimination, and click advertising, in 'ACMQueue', 2013, no. 11, p. 10 ff.; A. Datta, M.C. Tschantz, A. Datta, Automated experiments on ad privacy setting, in 'Proceedings on Privacy Enchancing Technologies', 2015, no. 1, p. 92 ff;

[37] A. Lambreckt, C. Tucker, *Algorithmic Bias? En Empirical Study into Apparent Gender-Based Discrimination in the Displey of STEM Career Ads*, in Management Science, 2019, p. 2966 ff.

not of diseases to decide[38] [39]. Examples that heighten concern and increase efforts to pursue the promised algorithmic objectivity and fairness.

On the other hand, one has to ask what ethics is meant by artificial intelligence and to whom is it really directed, whether to the AI itself or to the human being who—as we have seen with autonomous weapons—has to control it.

The comparison between human and artificial here constitutes a point of first interest in the twofold direction: to verify the differences (if any) between one and the other, to highlight what AI can tell us about human ethics.

Certainly, human ethics is inseparable from the human, within whose horizon action unfolds, and *the human being is an ontologically textualised entity.*

What does textualised entity mean? That narration is not just activity, the telling of stories, the construction of myths and legends, etc., but is an essential part—the 'non-technical' and non-serial reproducible essence[39]—of *homo viator* who is *homo incarnatus.* This implies that for the human it is not possible to speak of *hardware* and *software* by analogically speaking of the *body* and *functions of action* (if we do not want to speak of *spirit*, as in philosophical reflection since the Middle Ages); that for the human consciousness passes through self-consciousness, through the understanding of reality outside of oneself that leads to the understanding of the reality of, and within, oneself. It will suffice for me to recall—intending in this way to recall his thought in its entirety—the contribution that Paul Ricoeur [44–53] provides on the subject of the *homme fallible* and *capable,*[40] of narrative identity,[41] of linguistic hermeneutics and the age of secularization,[42] of social and political paradoxes,[43] of "une liberté seulement humaine",[44] as the well-known conclusion of the volume on the voluntary and involuntary that opens the youthful philosophy of the will[45] [21] titles.

The human universe is composed of fallibility, not of the predictable precision of computation, of a political, social and religious symbolic imaginary that animates and moves identity and social dynamics from the depths, for actions that in order

---

[38] Obermeyer Z., Emanuel E.J., *Predicting the Future. Big data, Machin learning and Clinical Medicine,* in The New England Journal of Medicine, 2016, no. 375, pp. 1216–1219.

[39] Cf. the well-known thesis of M. Heidegger, *Vortäge und Aufsätze,* in GA, Bd. 7, Frankfurt a.M., 2000, p. 5: "das Wesen der Tchnik ist überhaupt nicht technisch".

[40] P. Ricoeur, *Finitude et culpabilité*, Paris, 1960; *Parcours de la reconnaissance*, Paris, 2004.

[41] *Soi même comme un autre*, Paris, 1990.

[42] *Le conflit des interprétations*, Paris 1969; *Du texte à l'action*, Paris, 1986; *La religion pour penser*, Paris, 2021.

[43] *Philosophie, éthique et politique*, Paris, 2017; *Politique, économie et société*, Paris, 2019.

[44] *Le volontaire et l'involontaire*, Paris, 1950.

[45] A reference to Ricoeurian reflection with reference to AI can be found in the essay by G. Grandi, T Scantambulo, *Automatic Learning and Human Decision Making*, in F. Fossa, V. Schiaffonati, G. Tamburrini, eds, pp. 48–52; in this case, the reference is not to the narrative dimension of the human being, the *soi même comme un autre*, to use the philosopher's expression, recalled in these pages, but the three levels of attention for the ethical analysis of choice are applied to the issues concerning the development of AI.

to be understood must be textualised, in order to be hermeneutically grasped must therefore be narrated, to others and to oneself.

And before that, the question of the *body-proprio* appears to be diriment, according to Ricoeur, embodied in something that is not simply software: "ce sujet c'est moi *et* toi (…) Le corps propre est le corps de quelqu'un, le corps d'un sujet, mon corps et ton corps (…) la connaissance de la subjectivité ne se ré ré réalisée de un corps, mon corps et ton corps (…).) Le corps propre est le corps de quelqu'un, le corps d'un sujet, *mon* corps et *ton* corps (…) la connaissance de la subjectivité ne se réduit pas plus à l'introspection que la psychologie empirique ne se réduit à la psycologie de comportement".[46]

This is the basis for the possibility of ethical verification of human action.[47]

To think of an ethics of AI, on the other hand, means starting not from the textualisation but from the *computational basis of the algorithm*; it means thinking that one must separate the action from the agent with the first effect of *purging the concept of ethics from the concept of responsibility.* As Floridi writes, it is not by chance that 'whether an agent is really a moral agent depends on whether the agent can perform morally qualifiable actions, but not on whether the moral agent can or should be held morally responsible for such actions'.[48] The framework within which one must move in this sense is that, again illustrated by Floridi, for whom AI introduces a new form of action with the capacity not 'to reproduce human intelligence, but actually [with] the capacity to do without it'.[49]

This means that *the AI does not interpret reality but* modifies it, modifies it by simplifying and purging it, just as it simplifies action by reducing it to an event that takes place[50] [23–25], without the narrative, identity, consciousness apparatus that qualifies the human world.

Two areas may perhaps be illustrative.[51] Firstly, it is not the machine that works in place of the human being—as was also feared by some—but there is a risk—proven in recent years, especially post-Covid19—that it is the human being who works like a machine [2], as Antonio Casilli[52] [8] demonstrates very well. Having identified the basic criterion with that of algorithmic efficiency, typical of the *gig-economy*, human intelligence/human capacity[53] must be adapted to it, now absorbed in a system for

46 *Ibid*, p. 12.

47 On this basis I intend and apply, from the perspective of these pages, the distinction that Floridi makes between 'intelligent technologies' and 'truly intelligent technologies', L. Floridi, *Ethics of Artificial Intelligence,* cit., p. 50.

48 Ibid, p. 213.

49 Ibid, p. 52.

50 It is no coincidence that, in Martin Heidegger's philosophy, the event (*Erlebnis*) and avoidance is the antechamber to nihistic annihilation; cf. M. Heidegger, *Zur Bestimmung der Philosophie*, in GA, Bd. 56/57, Frankfurt a.M., 1987; but also *Grudprobleme der Phänomenologie*, in GA, Bd. 58, Frankfurt a.M., 1992, p. 117.

51 A. Aloisi, V. De Stafano, *Il tuo capo è un algoritmo*, Roma-Bari, 2020.

52 A.A. Casilli, *En attendant les robots. Enquete sur le travail du clic*, Paris, 2019.

53 Ibid, p. 97 ff.

which the term 'algocracy' is used, i.e. 'the networked digital environment in which power is exercised more and more by algorithms'[54] [13].

This leads to the second area, and to the afterthought behind the machinisation of the human being. The functioning of AI *does not have a semantic basis* but a *logical-formal basis of a statistical kind*[55] [62] requires a world that tends to be stable; in other words, by managing a mass of considerable data, it is able to connect data not by relating their content but by computing their form. Emblematically, Giovanni Sartor remarks that: "intelligence can be obtained by reproducing cerebral hardware (neurons) rather than by reproducing conscious thought processes (reasoning). Consequently, such research assumes that intelligent behaviour results primarily from flexible adaptation to the environment (with which all animals with a nervous system are endowed), rather than from reasoning'.[56] Add to this, as Gerd Gigerenzer notes[57], that AI computation is better in a stable world than in an unstable reality in which it struggles to make predictively accurate connections,[58] and thus the need arises to have a stable reality in which the AI can efficiently perform the task required of it. This means that reality must simplify itself as much as possible in order to acquire that stability necessary for its processualisation and, therefore, the means, the instrument, runs the risk of becoming an end, and thus of directing the very configuration of reality.

More so, then, than the principles of AI ethics, on which, as mentioned, a series of very interesting studies are flourishing, it is perhaps the case, as anticipated, to reflect preliminarily on the distinction between human and artificial in order to avoid a rapid re-ontologisation of the former into the latter. Not a transformation of the human into a machine, obviously, but a treating the human and its reality with the algorithmic logic of AI, losing its specificity and meaning. In these terms—to return to the three areas discussed in the previous paragraph—the very useful tool as a dating platform defines love with the *satisfaction of* profile compatibility, identifying profile and person; the very useful tool for improving the legal qualification of a circumstance or status on a broader casuistry than the one knowable by the jurist alone, ends up treating the singularity of the single case with the statistical computation of precedents; just as the autonomous weapon is called upon to decide on how to act to achieve the objective without considering the context in which the action is placed, necessarily, for an ethical (as well as legal, economic and political) evaluation.

A further example may represent the point even better. Currently, the best online automatic translator is rated DEEPL; its operation is simple and relies on a self-learning AI. Which is? It computes the match rate between individual words and, from time to time, chooses the translation that fits best. With a basic logic that belongs

[54] A. Delfanti, A. Arvidsson, *Introduction to Digital Media*, Bologna, 2013, p. 23.

[55] Cf. U. von Luxburg, B. Schölkopf, *Statistic Learning Theory*, in D.M. Gabbay, S. Hartmann, J. Woods eds, *Inductive Logic*, s.l., p. 651 ff.

[56] *Ibid*, p. 56.

[57] G. Gigerenzer, *How to Stay Smart in a Smart World. Wy Human Intelligence Still Beats Algorithms*, London, 2022.

[58] G. Sartor, *L'intelligenza artificiale e il diritto*, cit., p. 49 ff., clarifies the meaning of predictivity.

to all AIs: not to understand a text but to connect words that are understood as mere codes, like variables or symbols X, Y, Z. According to this logic, the more data, the more efficient and precise the result.

The human translator does not respect this logic, which is not that of the human being in his speaking or writing, in translating from one language to another; the human being communicates meaning content (and communicates himself every time he textualises). Ricoeur demonstrates this well when he presents precisely translation as a model of social communication[59]: "que signifie pouvoire traduire? Cette possibilité, ou plutôt cette capacité, n'est pas seulement vérifiée par le fait que l'on arrive effectivement à tarduire discours et textes d'une lange dans une autre, sans perte sémantique totalment dommageable et sourtout absolument irréparable. La possibilité de traduire est plus fondamentalmente postulée comme un a priori de la communicazion. En se sens je parlerai du 'principe d'universelle traducibilité'. La traduction est de fait, la traductibilité de droit'.[60]

The text, therefore, is something more than the transmission of a message, it is something that depends on the specific personality-identity-body of the textualising subject. The risk is that love is misunderstood with compatibility, that judgement is reduced to decision, that the undertaking of a military mission is misunderstood with the achievement of the objective, that translation is saying the same thing in a different language. It is the person of the lover in the relationship with the beloved that makes love something that cannot be defined[61] [28], it is the assessment of what lies behind the agent's behaviour that qualifies a defendant's conduct whether it is wilful or culpable[62] [34, 43], it is the ability to understand the situation that makes a military action ethically correct or not, it is the communication of the world of the author of the text and his very personal way of understanding it that makes a translation good or not.[63]

So, the ethics of AI seems, in some ways, the ethics to be always referred to the human being who has to control his technologically advanced instrument (as in the case of the autonomous weapon for which 'meaningful human control' applies), and thus respects classical ethics; in other ways, however, it seems to modify the very paradigm of ethics, calling for a split between action and history, between history and responsibility, and thus between action and responsibility.[64]

In fact, a major contribution that these recent studies provide is precisely to verify the classical paradigm of ethics, which is frequently questioned even outside the

---

[59] P. Ricoeur, *Politique, économie et société*, cit., p. 272.

[60] *Ibid*, p. 273.

[61] Cf. J. Lacan, *Le Séminaire, livre VIII. Le tranfert*, Paris, 2001.

[62] Cf. R. Rampioni, *Dalla parte degli "ingenui". Considerazioni in tema di tipicità, offesa e c.d. "giurisprudenza creativa"*, Padua, 2007, p. 154 ss.; V. Manes, *I recenti tracciati della giurisprudenza costituzionale in materia di offensività e ragionezza*, in "Diritto penale contemporaneo", 2012, n. 1, p. 99 ss.

[63] Cf. P. Ricoeur, *Sur la traduction*, Paris, 2004; *Du texte à l'action*, cit.

[64] In this case, however, while the legal issue of the evaluation of the action would remain fully operational and necessary, not only but also for compensation purposes, the need for an ethical evaluation of the action and with it an ethics itself would gradually be lost.

sphere of AI and as a way of rethinking human ethics itself. Indicative and representative of this more widespread interest in the paradigm shift is, deliberately identified not in ethics but in the legal sphere, represented in the science of criminal law; a current of thought understands the crime as the action abstractly envisaged as such, it is the so-called criminal law of the fact[65] [11, 26, 27, 42, 58, 59], and not as the history within which the action is inscribed with the consequence that the correspondence between concrete action and abstract action is sufficient for the deductive subsistence of the crime. A great Italian penalist, Francesco Carnelutti, icastically brings the example of the rule of homicide in the Italian penal code[66] and notes: 'A man is but a story. The brother plunges the dagger into his brother's chest. And before? And afterwards? Who was the slayer? And who is the slayer? Two stories: 'He who kills a man' [the incriminating norm] implies the whole problem of the past. And just as the notion of crime implies the problem of the past, so the notion of punishment opens up the problem of the future. Past and future. Everything. The whole of life. The whole of history'[67] [7]. The legally (in some ways, also ethically relevant) question is not whether only the subject has caused the death of another subject, in the case by stabbing him, the question is why? What came first in the relationship between subjects? What triggered this action? Without answering these questions, how can one judge?[68] [22, 57] And judging—going back to the judgement/decision distinction considered in the previous paragraph—does not only mean deciding whether the defendant is innocent or guilty, judging means in the first case (innocence) justifying freedom, in the second case (guilt) assessing the punishment to be applied (punishment varying from case to case within the range identified by the code).

If therefore, from the perspective of applying AI ethics to the people who control them, little new can be said in the light of the new technologies, remaining de facto an ethics directed at the human,[69] much needs to be thought about in the case of

[65] As contrasted in doctrine to the so-called author's criminal law. See G. Radbruch, *Faschistiches Strafrecht in Der Morgen*, 1933, p. 436 ff., now in *Gesamtausgabe*, Bd. 8, *Strafrecht*, II, edited by Arthur Kaufmann, 1998, Frankfurt A.m., p. 225 ff.; G. Spendel, *Zur Notwendigkeit des Objektivismus im Strafrecht*, in "Zeitschrift für die gesamte Strafrechtswissenschaft", 1953, p. 519 ff.; G. Dahm, *Der Tätertyp im Strafrecht*, Leipzig, 1940, spec. p. 15 ff.; F. Schaffstein, *Rechtswidrigkeit und Shuld im Aufbau des neuen Strafrechtssystems*, in "Zeitschrift für die gesamte Strafrechtswissenschaft", 1938, p. 300 ff.; more recently G. Jacobs, *Bürgerstrafrecht und Feindstrafrecht*, in "HRRS", 2004, p. 88 ff.; *Terroristen als Persone nim Recht?* in 'Zeitschrift für die gesamte Strafrechtswissenschaft', 2005, p. 839 ff.

[66] Article 575 of the criminal code: 'Whoever causes the death of a man shall be punished by imprisonment of not less than twenty-one years'.

[67] F. Carnelutti, *Art of Law*, Turin, 2017, p. 46.

[68] Cf. S. Satta, *Il mistero del processo*, Milan, 1994; in other respects, the considerations on 'emotional intelligence' and not only 'cognitive' by D. Goleman, *Emotional Intelligence*, New York, 1997, remain interesting in their application to the legal and procedural world.

[69] In this case, much remains to be done in terms of how, and the how is no small matter, as also seen above, to programmatically translate the established principles and rules so that they can also be followed by the AI.

rethinking reality in the light of the stability necessary for AI, and, beyond AI, of rethinking action in the light of its mere adaptation to the case.

In McCarthey's definition from which we move, not surprisingly, the yardstick is not action in itself and in its context, action is in fact measured by effect and only by effect, giving not only AI a qualification projected on effect and not on capacity, but also ethics a qualification distant from responsibility.

2.4 Conclusion: The Other Pinocchio and the Anthropocentric Overcoming.

Carlo Collodi thought—we are in 1883, the date of the wooden character's first appearance—of a 'puppet without strings' in which the absence of strings—actually but above all metaphorically—indicates full autonomy; Pinocchio does not actually respond to any design programme, he is absolutely autonomous in determining himself and in choosing, and, not by chance, he is always confronted with ethical dilemmas to which, for a good part of the book, he responds by preferring evil to good. Pinocchio is, à la Dostoevsky, sensitive to the 'taste for evil'. Then, at the end of the tale of his adventures, he turns into a real child. That is when the story of the puppet Pinocchio ends and that of the child Pinocchio begins.

But what differentiates Pinocchio between the 'puppet' and 'child' phases?

It is not an easy or intuitive question to answer. In another case—as literary as it is cinematic—*Bicentennial Man*[70] [4, 5]—the protagonist Andrew Martin initiates a legal procedure (of very long duration) for the recognition of his status as a human, which ends in actual recognition at the moment when, having acquired even the last human connotation, mortality, he is admitted among the circle of human beings, he who was born as a positronic android. The judgment of the World Council comes for Andrew *in limine mortis*, exactly at the moment when—further proving all his humanity—he dies. In this case, the machine's race to be recognised *as* human passes through a progressive rapprochement in the body (Ricoeur's body-proper) and in the way of seeing life (self-consciousness). One could say—modifying the mental experiment of Theseus' ship[71] [54]—that Martin becomes human when, by replacing his body piece by piece, he ends up corresponding in every way with a human being. Compared to this, AI, at least in its present state, is still a long way from coming remotely close to the human, and it is indeed 'digital utopians' who believe, in the short or at least medium term, that this ambition will come true.

The case of Pinocchio is different, as mentioned, and perhaps even more interesting to consider. He does not slowly become human through the modification

[70] Isaac Asimov's 1976 short story, *The Bicentennial Man,* continued in 1992 with *The Positronic Man,* written by Asimov and Robert Silverberg, had a film adaptation, directed by Chris Columbus, in 1999.

[71] On identity in the terms discussed here see C. Romano, *L'dentité humaine en dialogue*, Paris, 2022, pp. 65 ff. 217 ff.

of his hardware and software because from the outset he enjoys full autonomy and consciousness (both his own and mediated by the Talking Cricket). Pinocchio is born, even before he is born,[72] endowed with words, desires, ambitions. Word, desires and ambitions that he does not fail to use and put into play throughout his puppet phase; it is not therefore these characteristics proper to the human that are lacking to allow him the transformation into a real child, while for Andrew Martin it is precisely a progressive process of acquiring self-awareness that leads him to desire the recognition of humanity. And it is certainly not simple—indeed the literature is as varied as ever on the point—to understand the reasons that ultimately made him a flesh-and-blood child. The simplest solution[73] [6, 19] is the one that sees the transformation of the puppet into a child at the moment in which Pinocchio understands that one must want good, that he knows the sweetness of good and therefore stops choosing the taste of evil; and the sweetness of good is to be polite, altruistic, respectful, good citizens, in short, rather than embracing the selfishness of evil that embodies the bad citizen. A message that reflects the Risorgimento perspective in which Collodi writes and thinks.

I do not intend to dwell too much on this splendid and enigmatic literary character just yet, but it does seem, on the one hand, to be the goal to be achieved by those who aspire (the 'digital utopians') to an equalisation of man and machine, in other respects making man a true Creator; at the same time and on the other hand, Pinocchio also represents the proven proof that the 'techno-sceptics' can point to in order to demonstrate how the equalisation of status between human and machine can never be achieved.

In these pages we have not intended to move between these two poles, on the contrary, we have tried to indicate some points that we consider useful to keep under discussion because they are important both in the understanding of the technological innovations we are witnessing, and for what AI is able to tell us, backwards or *forwards*, with respect to human ethics.

The Pinocchio 3.0 of AI is certainly not like Collodi's original, it is *another Pinocchio* that perhaps prepares for the definitive departure from anthropocentrism—whether Collodi's dimension is fully anthropocentric as well as Asimov's is clear—for a *functional reduction of life* or for a *post-human turn*, as has been widely rumoured for some decades[74] [35, 41]; perhaps, and at the same time, it warns

---

[72] On closer inspection, the puppet has been speaking (in full self-consciousness) ever since the state of a simple log of wood and thus before its artificial transformation into the form of a puppet by Geppetto.

[73] When Collodi writes, it must be remembered, that it is the post-unification season in which the Italian state was born but, as Giacomo Biffi observes, 'a self-conscious nation was not born at all and agreed on the values that give meaning to life' (G. Biffi, *Contro Mastro Ciliegia*, Milan, 1977, p. 11); and in this scenario Collodi's book, as has been authoritatively specified by the philosopher of law Vittorio Frosini, "reveals itself (...) as a typically Risorgimento myth, at the sunset of an era" (V. Frosini, *La filosofia politica di Pinocchio*, Roma, 1980, p. 22).

[74] Cf. R. Pepperell, *The Posthuman Condition. Consciousness beyond the brain*, Portland, 1995; R. Marchesini, *Post-human*, Turin, 2002.

against rethinking human categories by applying them to the non-human—responsibility, guilt, freedom—for the risk of easily ending up transforming human life into the functionalisation of force, the one that ends up hierarchising all the others: that of power that prevails; or perhaps, again, it demonstrates the loneliness in creation that the human being feels and the need for, once again, another similar and equal creature to be born from his own rib, and this time by his own will and ability, bringing to fruition the modern atheism of *homo faber*[75] [14]. Certainly, the other Pinocchio 3.0 helps to better perceive that 'liberté seulement humane' of which Ricoeur speaks, inviting us to treasure it and not waste it, restoring meaning to ethics and ethics to life. It helps to understand how there is an ontological difference between human intelligence and artificial intelligence, which is built on *understanding* the history (from a narrative perspective) of the former and on *computing according to* classifications of the latter. After all, it is precisely the main requirement to operate from an AI perspective, the classification of reality, that constitutes its greatest limitation and point of departure from human action. This is made explicit in the essential distinction between the *narrative dimension*, which is proper to the human being, and the classificatory*(-defining) dimension*, which is necessary in AI.

That AI is a tool, increasingly useful and perhaps indispensable, does not even seem debatable; and that its presence will gradually become more and more evident in everyday life, this too does not seem to be open to question. And as a tool, devoid of ambitions that are more science fiction than scientific, it can certainly enrich and improve the quality of human action and the human condition and its ethicality, becoming a new and fascinating chapter in the story of human life on this Earth. Decisive in the coming years will be the recognition or otherwise of the distinction—marked not surprisingly in jurisprudence—between 'relying' on the instrument, blindly delegating decision-making to a blind system, or 'considering' the instrument as part of an activity of judgement and understanding of reality that remains within human availability.

It seems quite clear that, then, one of the choices waiting to be made will be whether to embrace a post-humanism (or trans-humanism, according to yet another perspective) that brings the human being functionally close to the machine or whether to continue with an anthropocentric perspective, which entails the challenge of bringing the machine as close as possible to the human being. Two specular but profoundly different perspectives that divergently interpret the meaning of *Etichs and Artifical intelligence.*

References

1. Aa. Vv.: A general reinforcement learning algorithm that master chess shogi and go through self-play. Science 362 (2018)
2. Aloisi, A., De Stafano, V.: Il tuo capo è un algoritmo. Roma-Bari (2020)
3. Arkin, R.C.: Governing Lethal Behavior in Autonomous Robots. Boca (2009)

[75] On this point, the pages of A. Del Noce, *Il problema dell'ateismo,* Bologna, 1964, remain valuable.

4. Asimov, I., Silverberg R.: The Positronic Man. London (1992)
5. Asimov, I.: The Bicentennal Man. New York (1976)
6. Biffi, G.: Contro Mastro Ciliegia. Milano (1977)
7. Carnelutti, F.: Arte del diritto. Torino (2017)
8. Casilli, A.A.: En attendant les robot. Enquete sur le travail du clic. Paris (2019)
9. Contissa, G., et al.: Claudette meets GDPR: automating the evaluation of privacy policies using artificial intelligence. SSRN-Electron. J. (2018)
10. Crootof, R.: A meaningful floor for "meaningful human control". Temple Int. Comp. Law J. 30, 53 ss (2016).
11. Dahm, G.: Der Tätertyp im Strafrecht. Leipzig (1940)
12. Datta, A., Tschantz, M.C., Datta, A.: Automated experiments on Ad privacy setting. In: Proceedings on Privacy Enchancing Technologies, vol. 1, p. 92 ss (2015).
13. Delfanti, A., Arvidsson, A.: Introduzione ai media digitali. Bologna (2013)
14. Del Noce A.: Il problema dell'ateismo, Bologna (1964)
15. Fazel, S., et al.: Preciction of violent reoffending in prisoners and individuals on probation. A Durtch validation study (OxRec). Sci. Rep. 9(841), 1 ss (2019)
16. Finkler, E.J, Eastwick, P.W., et al.: Online dating. A critical analysis from the perspective of psychological science. In: Psychological Science in the Public Interest, vol. 13 (2012)
17. Floridi, L.: Etica dell'intelligenza artificiale. Milano (2022)
18. Foti, A., Marino, D.: Blockchain and charities: a systemic opportunity to create social value. In: Marino, D., Monaca, M. (a cura di) Economic and Policy Implications of Artificial Intelligence. Studies in Systems, Decision and Control. Cham, pp. 145–148 (2022)
19. Frosini, V.: La filosofia politica di Pinocchio. Roma (1980)
20. Gigerenzer, G.: How to Stay Smart in a Smart World. Wy Human Intelligence Still Beats Algorithms. London (2022)
21. Grandi, G., Scantambulo, T.: Apprendimento automatico e decisione umana. In: Fossa, F., Schiaffonati, V., Tamburrini, G. (eds.) A cura di, pp. 29–53. Roma, Automi e persone (2021)
22. Goleman, D.: Emotional Intelligence. New York (1997)
23. Heidegger, M.: Zur Bestimmung der Philosophie. In: GA, Bd. 56/57, Frankfurt a.M. (1987)
24. Heidegger, M.: Grudprobleme der Phänomenologie, in GA, Bd. 58, Frankfurt a.M. (1992)
25. Heidegger, M.: Vortäge und Aufsätze, in GA, Bd. 7, Frankfurt a.M. (2000)
26. Jacobs, G.: Bürgerstrafrecht und Feindstrafrecht. HRRS (2004)
27. Jacobs, G.: Terroristen als Persone nim Recht? In: Zeitschrift für die gesamte Strafrechtswissenschaft, p. 839 ss (2005)
28. Lacan, J.: Le Séminaire, livre VIII. Le tranfert. Paris (2001)
29. Lagiola, F., Rovatti, R., Sartor, G.: Algorithmic fairness through group parities? The case of COMPAS-SAPMOC. AI Soc. 38, 462 ss (2023)
30. Lambreckt, A., Tucker, C.: Algoritmic bias? En empirical study into apparent gender-based discrimination in the display of STEM career ads. Manag. Sci. 2966 ss (2019)
31. Lazzini, F.: Etica digitale e intelligenza artificiale. Torino (2023)
32. Legg, S., Hutter, M.: A collection of intelligence. In: Goertzel, B., Wang, P. (a cura di) Advances in Artificial General Intelligence: Concepts, Architecture and Algorithms. Amsterdam (2007)
33. Lin, P., Jenkins, R., Abney, K.: Robot Etics 2.0. From Autonomous Cars to Artificial Intelligence. New York (2017)
34. Manes, V.: I recenti tracciati della giurisprudenza costituzionale in materia di offensività e ragionevolezza. In: Diritto penale contemporaneo. vol. 1, p. 99 ss (2012)
35. Marchesini, R.: Post-human. Torino (2002)
36. McCathey, J., Minsky, M.L., Rochester, N., Shannon, C.E.: A proposal for the Darthmouth summer research project on artificial intelligence, 31 Aug 1955. AI Mag 27 (2006)
37. Musiani, F.: Governance algoritmica: sorveglianza, censura e diritti fondamentali. In: Fossa, F., Schiaffonati, V., Tamburrini, G. (eds.) A cura di, pp. 95–113. Roma, Automi e persone (2021)
38. O'Neil, C.: Weapons of Math Destruction. How Big Data Increases Inequality and Threatens Democracy. New York (2016)

39. Obermeyer, Z., Emanuel, E.J.: Predicting the future. Big data, machin learning and clinical medicine. New England J. Med. 375, 1216–1219 (2016)
40. Oswald, M., et al.: Algorithmic risk assessment politicing models. Lessons from the Duhram HART model and 'esperimental' proportionality. Inf. Commun. Technol. Law 27, 223 ss (2018)
41. Pepperell, R.: The Posthuman Condition. Consciousness Beyond the Brain. Portland (1995)
42. Radbruch, G.: Faschistiches Strafrecht in Der Morgen, 1933, ora in Gesamtausgabe, Bd. 8, Strafrecht, II, a cura di Arthur Kaufmann, 1998, Frankfurt A.m.
43. Rampioni, R.: Dalla parte degli "ingenui". Considerazioni in tema di tipicità, offesa e c.d. "giurisprudenza creativa". Padova (2007)
44. Ricoeur, P.: Le volontaire et l'involontaire. Paris (1950)
45. Ricoeur, P.: Finitude et culpabilité. Paris (1960)
46. Ricoeur, P.: Le conflit des interprétations. Paris (1969)
47. Ricoeur, P.: Du texte à l'action. Paris (1986)
48. Ricoeur, P.: Soi même comme un autre. Paris (1990)
49. Ricoeur, P.: Parcours de la reconnaissance. Paris (2004)
50. Ricoeur, P.: Sur la traduction. Paris (2004)
51. Ricoeur, P.: Philosophie, éthique et politique. Paris (2017)
52. Ricoeur, P.: Politique, économie et société. Paris (2019)
53. Ricoeur, P.: La religion pour penser. Paris (2021)
54. Romano, C.: L'dentité humaine en dialogue. Paris (2022)
55. Ruggeri, F.: et al.: Detecting and explaining unfairness in consumer contracts through memory networks. Artific. Intell. Law 30, 59 ss (2022)
56. Sartor, G.: L'intelligenza artificiale e il diritto. Torino (2022)
57. Satta, S.: Il mistero del processo. Milano (1994)
58. Schaffstein, F.: Rechtswidrigkeit und Shuld im Aufbau des neuen Strafrechtssystems. In: Zeitschrift für die gesamte Strafrechtswissenschaft (1938)
59. Spendel, G. (1953) Zur Notwendigkeit des Objektivismus im Strafrecht. In: Zeitschrift für die gesamte Strafrechtswissenschaft (1953)
60. Sweeney, L.: Discrimination in online ad delivery. Google ads, black names and white names, racial discrimination, and click advertising. ACMQueue 11, 10 ss (2013)
61. Tamburrini, G.: Etica delle machine. Roma (2020)
62. Von Luxburg, U., Schölkopf, B.: Statistic learning theory. In: Gabbay, D.M., Hartmann, S., Woods, J. (a cura di) Inductive Logic, s. l, p. 651 ss

Chapter 3
Artificial Intelligence and Ethics Challenges of Technologies

Domenico Marino, Daniele Cananzi, and Filippo Aragona

Abstract While on the one hand the development of artificial intelligence is having an increasingly important impact not only on the economic system, but also on people's lives, opening up new scenarios, imposing new constraints, and giving rise to new opportunities, at the same time it is emerging as a further potential threat to humanity. We need to abandon an approach based on an optimistic prejudice towards artificial intelligence, which is often presented as a panacea for all evils, with thaumaturgic properties that can be extended to all problems and situations. Artificial intelligence is a tool that must be used with care and caution, also taking into account the possible risks involved, risks that will become all the greater the more powerful the calculation tools are. This is why the issue of 'Machine Ethics' is being forcefully raised at the scientific level.

3.1 Introduction

While on the one hand the development of artificial intelligence is having an increasingly important impact not only on the economic system, but also on people's lives, opening up new scenarios, imposing new constraints, and giving rise to new opportunities, at the same time it is emerging as a further potential threat to humanity. We need to abandon an approach based on an optimistic prejudice towards artificial intelligence, which is often presented as a panacea for all evils, with thaumaturgic properties that can be extended to all problems and situations. Artificial intelligence is a tool that must be used with care and caution, also taking into account the possible risks involved, risks that will become all the greater the more powerful the calculation tools are. This is why the issue of 'Machine Ethics' is being forcefully raised at the scientific level.

Artificial intelligence is essentially the ability of machines to make autonomous decisions as a synthesis of a large amount of information that is continuously processed. Man is entrusting artificial intelligence with increasingly relevant and difficult choices. An artificial intelligence that is trained to reduce the fuel consumption of an aircraft by correcting its flight attitude can also bring the aircraft to a

D. Marino et al., *Ethics and Artificial Intelligence*, Studies in Systems, Decision and Control 519, https://doi.org/10.1007/978-3-031-50902-5_3

standstill if certain conditions are met to achieve its goal. Intelligent weapon systems constitute a very sensitive ethical frontier. If a weapon can autonomously decide whether to hit a hospital, a bridge or a military installation what makes us think that it cannot autonomously decide to hit a hospital? Or how would an autonomously driven vehicle behave when faced with the choice between risking self-destruction in a dangerous manoeuvre or running over a child crossing the road? The tool will achieve as efficiently as possible the result that man has set for it. It is clear at this point how important from an economic point of view and what the implications for business are of considerations of the ethics of artificial intelligence.

Isaac Asimov, one of the most famous science fiction writers of the 20th century had foreseen this problem and tried to codify it through the three laws of robotics:

- **First law—a robot cannot harm a human being or allow a human being to be harmed.**
- **Second law—a robot must obey orders given to it by humans except in cases where those orders conflict with the first law.**
- **Third law—a robot must protect its own existence as long as that protection does not conflict with the first or second law.**

These principles contain intuitive wisdom, although they may be oversimplified, and are difficult to translate into concrete actions due to the innate anthropomorphism, which portrays machines in the image of human beings. Artificial intelligence basically consists of the ability of machines to make autonomous decisions by synthesising an enormous amount of information that is continuously processed. Increasingly complex and relevant decisions are delegated to artificial intelligence by human beings.

The modern idea that emerges from Asimov's insights leads us to consider robots as endowed with a peculiar personality that cannot be attributed to humans. The philosophical and legal concept of a 'non-human person' introduces a complex universe of ethical, moral and legal issues concerning the recognition of rights and responsibilities that go beyond the human sphere. This concept has generated heated debates, challenging the conventions of definitions of person and citizenship, and forcing society to deeply scrutinise its relationship with the natural world and the artificial life forms it has begun to generate.

Philosophically speaking, the idea of extending a 'personality' or at least certain prerogatives to non-human entities can be rooted in various ethical theories. For instance, the ethics of deep environmentalism holds that all life forms possess intrinsic value and deserve respect regardless of their usefulness to human beings. This could lead to considering non-human entities as bearers of rights, especially those related to life, integrity and a healthy environment. From a legal perspective, some nations and jurisdictions have begun to recognise some form of legal personality for non-human entities. For example, New Zealand has granted legal personhood status to the Whanganui River. In addition, the European Union has advanced regulations that could grant a form of 'electronic personality' to advanced artificial intelligences, giving them legal rights and responsibilities.

However, the idea of conferring personality or legal rights on non-human entities also poses intricate questions. How to balance the rights of non-human entities with those of human beings? How to establish legal responsibilities for an entity without biological consciousness or intentionality?

As long as the concept of non-human personality remains confined to animals or, at most, natural entities, the problems that arise remain relatively marginal, mainly within the sphere of compensation for damages, usually attributable to those with custodial duties, or in a balance of interests where human interests receive superior protection. In general, if the concept of non-human personality remains confined to living beings, no conflicts arise that would undermine our traditional free-market economic model. However, when this concept is extended to digital non-human entities, a number of issues arise concerning the role of 'non-human persons' within the context of market liberalism. These questions raise complex issues concerning the definition of economic actors and the limits of applying free market principles to non-human entities. In the context of market liberalism, which advocates economic freedom, private property, and minimal state intervention, considering 'non-human persons' as participants in economic transactions entails numerous challenges and considerations.

Several aspects to consider include:

1. Property rights and contracts: In the context of the free market, individuals can own property and enter into contracts. Extending the concept of personhood to non-human entities could imply that they have property rights and the ability to enter into contracts. This raises questions about how to handle transactions between non-human and human entities and how to enforce contracts with entities that may lack consciousness or intentionality.
2. Legal responsibility: In the free market, individuals are held responsible for their actions and may be subject to legal liability. Extending these principles to non-human entities would require reflection on how to assign responsibility and possible sanctions to entities potentially lacking the capacity to understand the implications of their actions.
3. Competition and competitive advantage: In the free market, enterprises and individuals compete for competitive advantage. The inclusion of non-human entities in the economic sphere raises questions about the justice of competition between actors with diverse characteristics and resources.
4. Impacts on inequality: The introduction of non-human entities as economic participants could affect the distribution of resources and wealth. This could have implications for economic inequality, requiring reflection on how to balance the interests of non-human entities with those of human beings, especially in a context where resources may be limited.
5. Regulation and state intervention: The expansion of the concept of the non-human personality within the free market could generate a greater need for regulation

and state intervention to ensure the proper functioning of the market and the protection of the interests involved.

3.2 Artificial Intelligence and Ethics: A Survey

The contributions of thought on ethics and artificial intelligence have been very numerous in recent years. It is, therefore, very difficult to make a summary that does justice to the complexity of the problem and its particular multidisciplinary aspects. Therefore, it seemed appropriate to focus, diversifying them, on two important areas, the scientific contributions on the one hand and the contributions that derive from the official documents of supranational organisations that are committed to giving clear guidelines and rules on these aspects.

Starting to consider scientific contributions, one of the first works that needs to be mentioned is that of Floridi and Sanders [1]. In this pioneering work, which will later form the basis of subsequent developments [2–4], a theme is highlighted that appears to be central: the differentiation between the concept of morality and responsibility. In particular, in computer ethics, psi can imagine an artificial agent having morals, but without necessarily displaying forms of free will or in concept of responsibility. A kind of 'mindless morality' is thus postulated. Allen et al. [5] starting from the concept of morality as a well-developed philosophical category from which a moral responsibility for actions taken derives, note that using this same scheme in machine ethics may give rise to new problems because in the case of artificial intelligence performing harmful or illegal actions, it is not clear who could be held responsible. Because, even if we accept Floridi's principle of the split between morality and responsibility, society as a whole cannot avoid finding someone responsible for the damage produced, if only for compensation purposes. In Allen et al. [6] a comprehensive survey of the main works on machine ethics is made, then, starting with an interesting taxonomy based on three dimensions: analysis in relation to the object (ethical theories) and non-technical aspects and technical aspects. Anderson et al. [7] poses the problem of the effects on ethical issues of the creation of machines that have the capacity to build other machines, In Anderson and Anderson [8] it is argued that it is necessary to discover the ethically relevant characteristics and corresponding duties present in the various possible actions that a system can perform by identifying decision-making principles for cases of conflict between these duties. Lauer [9] starting from such striking examples of unethical AI built and distributed around the world argues that AI ethics cannot exist without a broader ethical culture. Benjamins [10] argues that the ethicality of AI depends on the choices that organisations developing or applying AI can make, which can only lead to a more or less responsible use of AI Porter [11] emphasises that ethical principles of AI alone are not sufficient to ensure responsible use of AI in companies. Companies need strong and mandatory governance controls, including tools for process management and the creation of associated audit trails to enforce their principles. Floridi [4] identifies 5 core principles for defining benevolent AI. Four of them are core principles commonly used in bioethics: beneficence, non-maleficence, autonomy and justice. The fifth principle to be added is explicability, understood both as an answer to the question "how does it work?" and as an answer to the question "who is responsible for the way it works?".

Within the broader literature on the ethical implications of artificial intelligence, it is useful to devote a focus to the relationship between artificial intelligence and business ethics. In this context, we have selected a few recent contributions that we consider most significant in the context of what is becoming a very large literature.

We start with Russel et al. [12] who raise the issue of the ethical implications of the use of autonomous and intelligent weapon systems, explaining that international humanitarian law—which regulates attacks on human beings in times of war—could still be applicable in the event of a violation of the rules. Kriebitz and Lutge [13] highlight the ethical risks underlying artificial intelligence in relation to human rights, arguing the need for AI to ensure decisions that do not conflict with them. Munoko et al. [14], after considering that auditing firms are signalling the use of artificial intelligence (AI) in their auditing and advisory functions, at the same time gradually realising that unintended consequences may also occur. Baker-Brunnbauer [15] shows how complex and still partly unresolved the topic of AI ethics is from a management perspective. Digital technologies and trends, partly driven by AI, will have an immense impact on consumer behaviour, the economy, societies and other sectors. Digitalisation is therefore a powerful transformation for global players and monopolists, who have the opportunity to increase their market power. However, at times in spite of transparency and privacy protection.

Haenlein et al. [16] point out that teaching ethical behaviour in universities and schools is more crucial than ever for learning to work with AI and acquiring social and human skills. In Martin [17], they discuss whether developers have a responsibility for their algorithms that are subsequently used and what the normative basis of this responsibility is. Algorithms are an important factor in ethical decisions and influence the delegation of roles and responsibilities within these decisions. Therefore, companies should be responsible not only for the value of an algorithm, but also for the design of who does what within the algorithmic decision. In Martin and Parmar [18] it is stated that organisations. in order to avoid ethical errors, must build systems that help protect against preventable errors and recover from unpredictable ones. In Martin and Waldman [19] it is argued that so far, the literature on algorithmic responsibility has evaded a fundamentally empirical question that is important for ethics and business management: The paper seeks to identify under what circumstances, if any, algorithmic decision-making systems can be considered legitimate.

Alongside the contributions in the scientific literature, some important ones to give a normative and not only scientific and philosophical content to the ethics of artificial intelligence should be noted. The Asilomar conference[1] (2017) is one of the first attempts in this direction. In January 2017, more than 100 researchers and entrepreneurs gathered in Asilomar and drew up a list of 23 principles that should guide the development of artificial intelligence to ensure that it is safe, ethical and

[1] https://ai-ethics.com/2017/08/11/future-of-life-institute-2017-asilomar-conference/.

beneficial. Asilomar's comprehensive list of Artificial Intelligence Principles reads like an extended version of Isaac Asimov's famous Three Laws of Robotics. The 23 points are grouped into three areas: Research Issues, Ethics and Values and Long-term Issues. Also noteworthy is the 'White Paper on the Application of Artificial Intelligence to Public Administration'[2] (2018) drafted by the Italian Government's Task Force on Applications of Artificial Intelligence. This also includes the *Preliminary Study on Ethics of Artificial Intelligence*[3] conducted in 2019 by UNESCO with the aim of drafting a recommendation addressed to UN member states to provide ethical guidance for the development and application of *artificial intelligence* so that it is *human-centred* and oriented towards the protection and promotion of human rights and dignity.

In the European space, too, such a path has been considered with the launch of a *soft-law* attempt to determine ethical criteria for *artificial intelligence.* In 2019, in fact; the European Commission published the *Guidelines for* Trustworthy *Artificial Intelligence*[4] in which seven key points are indicated that must be respected in order to make the use of *artificial intelligence* ethically trustworthy: (1) *human agency and oversight*, (2) *technical robustness and safety*, (3) *privacy and data governance*, (4) *transparency*, (5) *diversity, non-discrimination and fairness*, (6) *environmental and societal well-being*, (7) *accountability*.

In 2020, the European Commission drafted and published the *White Paper on Artificial Intelligence*[5] in which certain parameters are set for an ethically sustainable development of this type of technology so that it is aimed at improving the well-being of citizens while respecting fundamental rights and freedoms. This was followed by the Artificial Intelligence Act[6] whose discussion started in 2022, which is an EU directive that on the basis of the risk-based logic typical of European legislation attempts to regulate the increasing development and placing on the market of products using artificial intelligence and the Artificial Liability Directive[7] that identifies the liability of companies that use harmful artificial intelligence.

On 14 June 2023, the European Union passed the Artificial Intelligence Act, which aims to regulate the responsible use of artificial intelligence to improve the well-being of citizens. By prohibiting the invasive and discriminatory use of AI systems, it will focus on unacceptable risks to fundamental rights and security. There will be bans on risky applications such as behaviour manipulation and social scoring, while high-risk systems, such as those for critical infrastructure and crucial decisions, will be regulated through data governance and transparency criteria. The European Parliament expands the list of risky systems to include biometric categorisation and emotional

---

[2] https://libro-bianco-ia.readthedocs.io/it/latest/.

[3] https://en.unesco.org/artificial-intelligence/ethics.

[4] https://ec.europa.eu/futurium/en/ai-alliance-consultation.1.html.

[5] https://ec.europa.eu/info/sites/default/files/commission-white-paper-artificial-intelligence-feb2020_en.pdf.

[6] https://eur-lex.europa.eu/legal-content/IT/TXT/?uri=CELEX%3A52021PC0206.

[7] https://ec.europa.eu/info/files/proposal-directive-adapting-non-contractual-civil-liability-rules-artificial-intelligence_en.

recognition. It will be possible to complain about decisions of high-risk systems. The AI Act also covers generative AI and basic models, imposing transparency and safeguards. The balance between regulation and innovation is addressed through research exemptions and open source licences, as well as regulatory sandboxes. The AI Act is the first global AI law, prompting other legislators like the US to develop similar regulations to promote the safe and ethical use of AI in the US as well.

3.3 The Risks of Unethical Artificial Intelligence

A relevant topic that is much discussed within the scientific community is the possibility of artificial intelligence behaving unethically, i.e. the possibility that artificial intelligence may be unethical either because it lends itself to being used unethically or because it makes unethical decisions [20, 21].

These two aspects are very important and very relevant because of the repercussions they have and can ultimately lead to violations of people's fundamental rights.

In order to analyse the concept of unethical use, we must make one consideration and distinction. Firstly, it can be safely stated that technology in itself is neither good nor bad. Rather, it is the use made of a technology that makes it good or bad. Thus a drone can be used either to disperse fertiliser surgically or to kill by surgically bombing the enemy. But this is only partly true, because for some technologies and some technological products it is really hard to think of a good use. While it is true that dynamite can also be used for peaceful purposes, for example, in the controlled demolition of infrastructures, it is also true that its peaceful uses are certainly residual compared to its non-peaceful uses, and it is undeniable that there are also technologies and technological products for which it is very difficult, if not impossible, to think of a peaceful and ethical use. I challenge anyone to identify an ethical use for anti-personnel mines or a Kalashnikov.

Returning to artificial intelligence, we can outline some possible impacts that could lead to the use of artificial intelligence being classified as unethical. These certainly include those algorithms aimed at facial recognition, localisation and surveillance and which can be used as a tool for repressing freedoms or as a means of reducing and controlling individual freedoms by totalitarian regimes. Alongside these, there is certainly among the unethical uses the research to develop smart weapons, weapons capable of autonomously choosing and hitting the target. Finally, there are tools for distorting reality and orienting consensus and behaviour, and these are the generators of fake news, the trolls, the tools for creating distorted and fraudulent representations of rumours and images. If facial recognition, localisation and surveillance techniques still lend themselves to the possibility of 'dual use', for the other types of use of artificial intelligence the classification can only be 'totally unethical use'.

But alongside the unethical uses of artificial intelligence, there is also the not insignificant problem of unethical decisions that can be made by an artificial intelligence within ethically acceptable uses of these algorithms.

These unethical decisions can result either from the inability of the artificial intelligence to ethically discriminate the proposed strategies, or from the fact that like humans, artificial intelligences can develop biases, which can arise autonomously in the learning process, or can be induced by the proposed learning dataset. In the first case, deep learning algorithms are not super intelligences, but are more like children learning what you are trying to teach them, distorting the meaning and content and including things that were not part of what they were supposed to be taught, which in pedagogical terms we could call overfitting.

In the second case, however, the bias of the algorithm reflects the bias of the human being who constructed the data set for learning.

The problem of unethical decisions by artificial intelligence is relevant when one considers that these algorithms are beginning to be widely used in fields where unethical decisions can lead to very serious consequences. In the United States, the police use facial recognition software to identify the perpetrators of crimes. It can, therefore, happen that one is arrested and put in jail because of an erroneous recognition, or rather an erroneous association of the image of the criminal with the real person. Also in the United States, the software that was used to predict the likelihood of re-offending by criminals had developed a racial bias and, therefore, for the simple fact of not being of white race one was certain to face a heavier sentence. The algorithms that manage rider deliveries may penalise some workers and benefit others based on criteria that may be totally unethical and turn the algorithm into a kind of digital corporator.

In the case of the development of biases by algorithms, one can intervene by doing a data cleaning job to create bias-neutral datasets. But can the decision to eliminate biases inherent in the data not itself result from a bias?

In the case of the possibility of unethical decisions by artificial intelligence, we intervene by trying to insert some shared ethical references into the decisions. These references could be fairness, reasonableness, responsibility, safety and protection of privacy. However, the attempt to construct a moral philosophy for the machine is arduous and difficult, bearing in mind that it has taken man thousands of years to do this, and it is not certain that he has succeeded to the full.

3.4 Towards a Machine Ethics

It appears, in this light, necessary to start building the foundations of a 'Machine Ethics' that starts from the assumption that decisions made by artificial intelligence must take into account social values, moral and ethical considerations. The concept of 'Machine Ethics' is closely connected with the concept of Responsible Artificial Intelligence in the sense that decisions taken by artificial intelligence must be in line with the value system of society in order to ensure a fair and sustainable welfare.

And this is the most delicate point, because it is necessary to clearly define what the term Responsible means in concrete terms. It is not a question of imagining a responsibility that is linked to free will and the ability to discern between good and evil, which in ethically complex cases can only be resolved at the level of conscience. Conscience, which is and will probably always remain the dividing line between the human being and the machine, although it cannot be ruled out that in the future artificial super-intelligences will overcome these Pillars of Hercules. It is therefore a question of finding tools and methods to ensure that the choice of artificial intelligence does not deviate from what is universally accepted and acceptable human behaviour.

Virginia Dignum[8] identifies three levels of interaction between ethics and artificial intelligence:

- Ethics *by* Design, which consists of the technical/algorithmic integration of ethical reasoning capabilities into the behaviour of artificial intelligence and is realised through an approach that includes formal tools for describing a situation and ethical principles capable of autonomously making a judgement and justifying why a given decision is ethically, or not, acceptable.
- Ethics in Design, which identifies the normative and engineering methods that support the analysis and evaluation of the ethical implications of AI systems as they complement or replace traditional social structures, which means that the algorithms that govern our lives must be transparent, fair and accountable to the shared values of stakeholders.
- Ethics *per* Design which consists of designing codes of conduct, setting standards and certification processes that guarantee the integrity of developers and users as they research, design, build, deploy and operate artificial intelligent systems. Obviously in this case, designers must be given a clear and consistent set of ethical principles so that the system can make decisions according to these principles and warn users when it deviates from this value system.

As can be deduced, this involves implementing complex models whose actual functioning must be tested on a case-by-case basis. In general, when the decision-making activity of artificial intelligence may present risk profiles, it is always advisable to act in accordance with the precautionary principle. In such cases, therefore, the machine, however endowed with artificial intelligence, must **always** be **supervised by man.**

The principle of **human supervision over artificial intelligence decisions** must and will be increasingly safeguarded because only the synergy between artificial intelligence and human supervision can ensure an almost normal level of functionality even in the most serious cases.

This procedure called extended artificial intelligence makes it possible to associate human ethical drivers with the learning capabilities of artificial intelligences, thus avoiding both discriminatory bias and also avoiding the possibility that the decision taken in the restricted domain of artificial intelligence is optimal for a single problem,

---

[8] Dignum, Virginia. "Responsibility and Artificial Intelligence." *The Oxford Handbook of Ethics of AI* 4698 (2020): 215.

but devastating for the overall governance of the process. These ethical drivers must also be conveyed to artificial intelligences, assuming a human-artificial intelligence interaction in which humans always dictate the rules of the game and the machine is a useful problem-solving tool.

In essence, it is necessary to ensure that the human being is always able to intervene to ensure that the behaviour of Artificial Intelligence systems is always beneficial to mankind. It is therefore a matter of inserting a kind of emergency button that allows the action to be interrupted and the machine to be switched off, preventing the system from learning how to bypass this action should the observer find violations of predefined ethical principles.

In order to be ethical, artificial intelligence must reproduce a reference value pattern, based on those shared by human social and institutional structures. It is clear that when we speak of shared social and institutional structures, a certain ambiguity can be generated because different social systems today develop rules of social behaviour that can be very different. In general, therefore, we could define the ethical action of artificial intelligence as the ability to make decisions that are always capable of safeguarding the dignity of the person, operate according to the principles of subsidiarity and act in solidarity. This enormous power has laid the foundations for the creation of an ethics of technology that must take into account the **precautionary principle** when dealing with these extremely dangerous areas, the **principle of fairness** and equal opportunities, i.e. ensuring that **technological developments are inclusive, decrease inequalities** and also take into account procedural fairness, and the **principle of maximum social welfare,** i.e. that a technological development must correspond to a maximisation of social welfare.

The ethics of technology in today's world characterised by the increasing weight of artificial intelligence must evolve towards a trans-ethics of technology that must add new pillars to those previously set out that formed the structure of the ethics of technology. **The new values are the high reputational score that must characterise actions and individuals and transparency in the sense that the structure of algorithms and processes must be known to all those affected by the decision-making process.**

These ethical drivers must also be conveyed to artificial intelligences, assuming a human-artificial intelligence interaction in which man always dictates the rules of the game and the machine is a useful problem-solving tool. **The technological tool, such as a machine, is ethically neutral. It is neither good nor bad. In** order to avoid risks, it is therefore necessary that the machine, however endowed with artificial intelligence, is **always supervised by man** because, for example, an artificial intelligence will never be able to grasp, on the basis of the observations derived from its sensor system, the substantial difference between a bomb and fireworks. **Man–machine interaction will obviously become more complex as the computing capabilities and potential of** artificial intelligence become **greater than they are today,** when the artificial super intelligences whose capabilities we cannot even guess at today will be born. This is why it is necessary to start building a solid foundation for true machine ethics.

The 23 principles of Asilomar that characterise so-called beneficial artificial intelligence are a good starting point for this goal. However, to fully achieve the goal of beneficial artificial intelligence, we need to act on a more operational level. We must therefore start with the databases through which AI is trained. These databases must be ethical in the sense that not only must they be structured to avoid the emergence of bias (e.g. racist bias), but they must contain virtuous patterns of behaviour that must be assimilated by the AI. A 'pedagogical' approach must therefore be used to ethically wean the AI, just as parents do with children when educating them to discern right from wrong. Machine Ethics' therefore needs to implement beliefs and convictions within the AI through a process of ethical constructivism that enables the AI to make decisions respecting the ethical values unanimously shared by mankind, rejecting behaviour that could undermine the AI's main purpose, which is to be beneficial to Humanity.

3.5 Implications for Business Ethics

The above considerations have considerable relevance for understanding what decisions must be taken by companies, what values companies must refer to and what alerts companies must consider when artificial intelligence enters their business in any capacity. In the applications of artificial intelligence, as we have seen, the concept of responsibility still remains divorced from autonomy in terms of choice. This means that a company could be held liable for the damage that an algorithm or an android may cause with their actions and decisions. As one can easily deduce, the scope of application potentially giving rise to liability issues are very wide and may even touch the criminal sphere and not only the civil one. Just to mention a few cases, a health robot could be called to answer for the damage potentially caused to a patient's health, a personnel selection algorithm could be called to answer for discrimination during selection, an algorithm used to diagnose illnesses could be called to answer for a wrong diagnosis, an autonomous driving car could be called to answer for liability connected with an accident, without considering all the controversies that could arise connected with the training of algorithms in relation to transparency and the protection of privacy. To minimise these risks, mechanisms need to be introduced that can at least partially limit the profile of corporate liability with respect to the negative effects that could arise from the autonomous choices of artificial intelligence. Firstly, consideration should be given to ensuring human monitoring of the decisions taken by artificial intelligence algorithms. For example, in the case of health diagnoses, it would in any case be necessary to ensure that a health professional agrees with the diagnosis obtained by the artificial intelligence, while in the case of autonomous driving, it would in any case be necessary to ensure that there is some form of human supervision of the behaviour of the artificial intelligence, with a human being able to take control of the driving at any time. This may of course entail a higher cost, but it may prevent worse damage. It is therefore a question of never allowing artificial intelligence to be used without control, with some form of human supervision at all times. In the medical field, this means that artificial intelligence diagnoses should be reduced to mere indications that need to be validated by a doctor. In the case of indications arising during mass screening, consider these as increased risk profiles for a certain group of patients in relation to a particular pathology, an increased risk that must then be managed by medical personnel.

The other aspect necessary to limit the risk of negative consequences in terms of liability is to ensure the appropriate transparency of the algorithms. Those who, to some extent, are subjected to the decisions of artificial intelligence must have at their disposal all the information on the training data, the structure of the algorithm and the operating procedures, so that they can exercise a form of control over the decisions, to be able to defend themselves effectively in the case of decisions that are deemed incorrect, thus exercising their right to an adversarial debate with the person, in this case a machine, who makes decisions that may affect the individual personal sphere.

But attention must also be paid to the problem of data selection and the learning process by which the algorithm is trained. For example, covid patient information, including medical scans, was collected and shared during the pandemic from the daily practice of the wards, which became enormously complex to manage at certain times during the peak of the epidemic. Researchers wanted to acquire data quickly, but this meant aggregating data from different observations or often data from unknown sources. As an example, they unintentionally used datasets that contained lung scans of children who did not have covids as archetypes of non-covid cases. But as a result, the artificial intelligence algorithm learned to identify the children, not the covid. To get correct results one must therefore have a certified input both with respect to the training the algorithm must have and with respect to the ultimate goal of the training. It is also, of course, necessary that the training of the algorithm is done correctly, even when it is continuously fed with new data. These are two conditions necessary to guarantee the goodness of an algorithm's response, but not yet sufficient to guarantee an ethical response. To guarantee an ethical response, the algorithm must be weaned and educated to make it give the ethically correct response, i.e. the one that reflects a shared human value system. This is why it would be appropriate to regulate a system for certifying the ethical decisions of an artificial intelligence algorithm. Only artificial intelligences capable of passing ethical decision tests should be certified and at least partially relieve the company using them of responsibility for any harmful decisions that might be made by them. Turing tests of artificial intelligence's ethical solutions should therefore be designed in such a way as to guarantee that their behaviour will not differ from that of a human obeying a unanimously shared value system. This could be a very important first regulatory step, a first attempt to give certain rules to an area that is growing rapidly and tumultuously and in which humans are still in many ways groping their way.

3.6 AI, Data and Privacy

A final aspect that needs to be addressed is the relationship between AI, data and privacy, as the significant increase in digital data that permeates every aspect of our lives, both inside and outside the medical field, offers multiple opportunities to monitor our health, understand the origins of diseases, develop new therapies and influence our lifestyle habits. At present, artificial intelligence (AI) is progressively penetrating the healthcare sector, demonstrating success in diagnosis through scanning and imaging, the creation or discovery of innovative therapies, patient selection, treatment planning and health monitoring. Furthermore, the use of AI presents opportunities to improve the administrative efficiency of the sector. However, the actual application of such technologies remains limited and the revolutionary advances that could transform the field of health are yet to be realised. The expanding frontiers of data that can be used to infer individual health status, as well as who can access it, raises concerns about our autonomy and privacy. An example of this is the use of health data by pharmaceutical and technology companies for commercial purposes. Moreover, health data are often plagued by imperfections and biases. When used to develop artificial intelligence systems, they can perpetuate discrimination and exacerbate inequalities. Another major concern is how technological advances contribute to national and international inequalities.

. The AI Act, mentioned above, in relation to the issue of privacy places great emphasis on safeguarding personal data and protecting the privacy of citizens. The adopted document contains specific provisions to ensure the secure processing of health data in accordance with privacy laws. It is recognised that health-related data are particularly sensitive and require a high level of protection. The AI Act lays down clear rules regarding the collection, processing and use of health data in the context of artificial intelligence systems. In accordance with the principle of data minimisation, the AI Act requires that the minimum amount of data necessary to achieve the intended purposes be used. This means that companies and healthcare providers must limit the collection and processing of health data to only what is strictly necessary for specific purposes such as diagnosis, treatment or medical research. Furthermore, the AI Act emphasises the importance of obtaining informed consent from patients for the processing of their health data. Citizens have the right to be adequately informed about how their data will be used and must give their explicit consent to the processing of this information. The proposed legislation also aims to ensure transparency in the use of artificial intelligence in the field of health. Organisations employing AI systems for health purposes must be transparent about how these systems work, including the logic behind decisions and the criteria used to make them. Finally, the AI Act provides for adequate security measures to protect health data from unauthorised access, loss or misuse. Organisations must take appropriate technical and organisational measures to ensure the security and integrity of health data, preventing any risk of privacy breaches.

This regulation has a strong impact on the development of cloud technologies for healthcare data. Migrating a hospital or healthcare company to the cloud is, in fact, a complex process that requires careful planning and implementation to ensure the success of the project. There are many opportunities that come with cloud migration, but there are also risks and problems that must be managed. Firstly, one of the main benefits of cloud migration for a hospital is the ability to reduce the cost of managing the IT infrastructure. By migrating to the cloud, the hospital can avoid having to manage and maintain an expensive and complex hardware infrastructure, as well as reduce the costs of IT personnel needed to maintain the hardware. In addition, using the cloud can offer greater flexibility and scalability, allowing the hospital to expand or reduce IT capacity quickly and efficiently. However, there are also some risks and problems associated with migrating healthcare facilities to the cloud. One of the main risks concerns data security and regulatory compliance. Healthcare facilities must ensure that sensitive patient data is adequately protected and that privacy and data protection laws are respected. In addition, migration to the cloud could lead to increased exposure to cybersecurity risks, such as phishing or malware attacks, which could compromise data security. Another issue could be the need to adapt existing applications to the cloud environment, which may require additional time and resources. Furthermore, the hospital must carefully assess its data security and privacy needs to determine the most appropriate cloud solution and ensure that the service is reliable, scalable and continuously available.

More generally, the AI Act also includes relevant ethical provisions to promote responsible and ethical use of AI. These provisions reflect an awareness of the importance of addressing the ethical issues associated with the widespread adoption of AI in various sectors, including the health sector, and recognise that AI must be developed and used with respect for fundamental values, human rights and fundamental freedoms. Preventing the harmful or discriminatory use of AI and promoting its proper application.

References

1. Floridi, L., Sanders, J.W.: On the morality of artificial agents. Minds Mach. **14**, 349–379 (2004)
2. Burr, C., Taddeo, M., Floridi, L.: The ethics of digital well-being: a thematic review. Sci. Eng. Ethics **26**, 2313–2343 (2020). ISSN 1353-3452
3. Floridi, et al.: What is data ethics? Philosoph. Trans. A (2016). ISSN 1471-2962
4. Floridi, L.: Translating principles into practices of digital ethics: five risks of being unethical. Philos. Technol. **32**, 185–193 (2019). ISSN 2210-5433
5. Allen, et al.: Why machine ethics? IEEE Intell. Syst. (2006). ISSN 1541-1672
6. Allen, C., et al: Artificial morality: top-down, bottom-up, and hybrid approaches. Ethics Inf. Technol. (2005). ISSN
7. Anderson, S.L., et al.: Machine Ethics. Cambridge University Press (2011). ISBN 9780521112352
8. Anderson, S.L., Anderson, M.: AI and ethics. AI Ethics **1**, 27–31 (2021). ISSN 2730-5961
9. Lauer, D.: You cannot have AI ethics without ethics. AI Ethics **1**, 21–25 (2021). ISSN 2730–5961
10. Benjamins, R.: A choices framework for the responsible use of AI. AI Ethics **1**, 49–53 (2021). ISSN 2730-5961

11. Eitel-Porter, R.: Beyond the promise: implementing ethical AI. AI Ethics **1**, 73–80 (2021). ISSN 2730-5961
12. Russell, S., Hauert, S., Altman, R., Veloso, M.: Robotics: ethics of artificial intelligence. Nature **521**, 415–418 (2015). ISSN 0028-0836
13. Kriebitz, A., Lutge, C.: Artificial intelligence and human rights: a business ethical assessment. Bus. Human Rights J. **5**(1), 84–104 (2020). ISSN 2057-0198
14. Munoko, I., Brown-Liburd, H.L., Vasarhelyi, M.: The ethical implications of using artificial intelligence in auditing. J. Bus. Ethics **167**, 209–234(2020). ISSN 0167-4544
15. Baker-Brunnbauer, J.: Management perspective of ethics in artificial intelligence. AI Ethics **1**, 173–181 (2021). ISSN 2730-5953
16. Haenlein, M., Huang, M.H., Kaplan, A.: Guest editorial: business ethics in the era of artificial intelligence. J. Bus. Ethics (2022). ISSN 0167-4544
17. Martin, K.: Ethical implications and accountability of algorithms. J. Bus. Ethics (2019). ISSN 0167-4544
18. Martin, K., Parmar, B.: Designing ethical technology requires systems for anticipation and resilience. MIT Sloan Manag. Rev. (2021). ISSN 1532-9194
19. Martin, K., Waldman A.: Are algorithmic decisions perceived as legitimate? The effect of process and outcomes on perceptions of legitimacy of algorithmic decisions. J. Bus. Ethics (2022). ISSN 0167–4544
20. Aragona, F., Marino, D.: Intelligenza artificiale ed Etica. Il Ponte **78**, 98–105 (2022). ISSN: 0032-423X
21. Marino: Intelligenza artificiale, Rubbettino (2020). ISBN 978-8849861624

Chapter 4
Case Studies

Domenico Marino, Daniele Cananzi, and Filippo Aragona

4.1 AI, Ethics and Health

4.1.1 IBM and Digital Health

IBM is not a company born in the twenty-first century like so many start-ups in the industry. It was born almost 150 years ago and as its first objective it had the production of tabulating machines, which were the distant ancestors of our computers and which functioned by inserting punched cards that were the device on which information was stored. The first name of this company was **Computing Tabulating Recording** (CTR), and only in 1924, when Thomas Watson was managing director, did it change its name to **International Business Machine (IBM). IBM then became during the so-called third industrial revolution, starting in the 1970s, a technology leader with its computers that were hailed as synonymous with quality and performance. However, the company on the wave of success became too big and elephantine with around 400,000 employees and went through an initial economic crisis in the second half of the 1980s that resulted in a major corporate restructuring through which, however, the company was able to return to profit after a few years. In 2005, IBM decided to exit the personal computer market, which it sold to the Chinese Lenovo, in order to devote itself to information technology and the development of supercomputers. As a result of this decision,** an innovative program called Watson was launched in 2006. Its name was not inspired by Sherlock Holmes' famous collaborator, but was given to honor the first of the brand's founder, who, as we have seen above, was called Watson. The current situation of IBM and its latest economic-managerial choices can be a very interesting case study and very rich in implications in relation to the most profitable and efficient strategies that companies operating in the field of advanced computing and artificial intelligence have to follow, and also of the ethical problems they have to face.

D. Marino et al., *Ethics and Artificial Intelligence*, Studies in Systems, Decision and Control 519, https://doi.org/10.1007/978-3-031-50902-5_4

Watson is a *question* answering *computing system*, i.e. a computer which, using artificial intelligence, is able to answer questions asked by humans directly. The progenitor of this class and computer became famous because he managed to beat the contestants of a quiz program in 2010.

In 2013, Watson Health was tested, which was supposed to be a computer capable of giving advanced medical answers, and with this application IBM managed last year to achieve $1 billion in sales. IBM's economic situation, however, has not been particularly prosperous in the last three years. Turnover decreased by about 8 per cent between 2018 and 2020 and by more than 25 per cent since 2010, to about $73 billion. EBITDA plummeted by almost 25% between 2018 and 2020 and profit fell by more than 40% over the same period. Therefore, at the end of 2020, IBM faced a new restructuring and decided to spin off its Information Technology business, which was worth about $20 billion in revenue. IBM therefore chose to focus on its core business, divesting those areas that were less profitable. As part of this corporate restructuring, it was also decided to spin off Watson Health.

The historical path of IBM's business organisation is very important to study, because it highlights certain aspects that allow us to understand certain mistakes that must be avoided in the management of advanced technologies and the application of AI to healthcare.

The first consideration that seems appropriate to make is that some applications, such as those of *Question Answering Computing Systems,* are commercially unattractive in perspective, all the more so when applied to healthcare.

It is quite obvious that the idea of voice interaction with a computer or an android has always been the secret dream of all programme developers and hardware manufacturers. In the medical field, the idea of developing an artificial intelligence that would respond vocally to scientific questions posed, as the medical hologram did in Star Trek, may seem appealing. But when we abandon science fiction and return to reality, we realise that voice interaction with the computer or algorithm can only be truly useful in a few limited cases. The case of disability management is one of them, but in most cases this excess of technology turns out to be an unnecessary waste, because it commits computing resources to the management of the voice interface and because the level of efficiency and effectiveness of the voice command is not always optimal.

In the healthcare sector, then, accessing and managing patient data is extremely complex. In order to function well, AI needs a huge amount of health data that is difficult and costly to store, not least because of the lack of predefined standards for collection and storage. In order to 'feed' our algorithm, therefore, we have to employ considerable resources.

But it is also the type of healthcare use of this voice medical interface that manifests many limitations. Watson Health was born with the claim of being able to provide an answer in terms of treatment for all major medical problems and, in particular, it was supposed to be able to provide support in the field of oncology.

Today, the technological capabilities of artificial intelligence algorithms are not yet able to achieve a level of efficiency in their responses that can replace a flesh-and-blood doctor. And this is the reason for the lack of success of Watson Health, which was supposed to be an innovative service, but which, in reality, had a level of operation that placed it below expectations.

In the medical field, artificial intelligence can be applied to numerous problems with better results than humans. In imaging diagnostics, in predicting outcomes, in simulating the dynamics of an epidemic, in mass screening and telemedicine, in analysing different therapeutic uses for existing drugs, in particular applications for the production of vaccines, but they cannot today replace a flesh-and-blood doctor in treating the real patient. AI applications in healthcare must, therefore, be supervised by humans and, therefore, these technologies, far from being autonomous decision-making systems, are rather supports for the decisions to be taken by healthcare personnel.

If, then, in the field of advanced research, futuristic solutions are to be sought and studied, in the commercial utilisation of discoveries and inventions, reference must be made to a concept of useful technology, even before innovative technology.

The IBM case we have described shows us how a gap exists today between research advances and commercial applications in the field of artificial intelligence.

An ambitious and visionary project from a scientific point of view turned out to be of little commercial success. This does not mean that we should not do advanced research, but rather that we should separate the areas of advanced research from those of commercial applications, because today's research will give rise to discoveries and inventions that we will be able to translate many years later into commercial products and are, therefore, an investment to be sustained. Useful technology is that which addresses real problems in an efficient manner, and it is that which must be linked to commercial applications and which can translate into an even substantial increase in the company's budget profits.

The strategic choice to focus on *Question Answering Computing Systems* proved to be inefficient from a business point of view for IBM, but the know-how developed constitutes a value that will probably produce its effects in ten or fifteen years' time in terms of the commercial development of new products. Wanting to commercially exploit advanced research products is a complex operation and almost never profitable in the short term.

4.1.2 AI and Health Data

One of the main problems when approaching artificial intelligence in general, and in the health sector in particular, is to understand its limits and to admit the possibility that errors may emerge from the processing. Too often, deep learning software is emphasised and its capabilities overestimated.

To properly govern the predictive capabilities of artificial intelligence, data cannot be ignored. The lesson that any first-year physics student must learn, if he wants to continue his course successfully and have a good chance of graduating, is the sacredness of data. It is the data of the experiment that is the material that can support all subsequent considerations, and if the data is not good, the whole castle that is built upon it easily falls down. This lesson is not always present when trying to apply artificial intelligence to a wide variety of fields. The ease of use of the tool can mislead one into thinking that it is enough to have large amounts of data to achieve a good result through deep learning. It is the power of the tool that is misleading and leads one to believe that artificial intelligence resembles a small box into which one only has to insert data to get a good result. This false belief became widespread in the scientific community, especially in the medical field, during the Covid 19 pandemic. An extremely large number of algorithms were developed that aimed to predict aspects of the Covid 19 pandemic in order to attempt to combat it effectively. In retrospect, however, what emerges, unfortunately, is that of the hundreds or thousands of artificial intelligence tools designed to better manage the pandemic, those that actually succeeded in passing the clinical test, i.e. the concrete impact with hospital reality, can be counted on half the fingers of one hand. This means that the cyclopean effort to develop solutions eventually gave birth to a mouse. Almost all applications had zero impact with respect to the aim of improving pandemic management. Algorithms for predicting the probability of worsening patients failed, early and rapid diagnosis programmes failed, and models for predicting the spread of the epidemic failed. Another well-known fact is that algorithms can develop biases, i.e. false representations that can affect the final result. For example, an Ai algorithm that estimates the probability of recidivism of a convicted offender may develop a racial bias by assigning a lower probability of recidivism to a white offender.

The origin of these failures lies in the quality of the data that is used to train the algorithms. Machine learning is often carried out lightly with unchecked, or unverified, or poor quality data. A poor quality database is already the first step towards certain failure. As I said before, physicists are taught early on the sacredness of data in experiments. The scientist becomes the priest who transforms the observation of reality and its measurement into data that must faithfully reproduce its characteristics. The study of data quality and error conditions thus becomes the prerequisite for any further processing. In the case of Covid, this has never happened, either because of the anxiety to have instruments capable of fighting this disease, or because of the lack of databases containing homogenous and certified data.

Frankenstein' databases were created from the aggregation of different databases that were not homogeneous and, above all, consisted of data collected with different methodologies and tools. Many tools, moreover, were developed by AI researchers who lacked the medical skills to detect flaws in the data or by medical researchers who lacked the mathematical skills to handle the problems of data collection.

If we want, therefore, to avoid the failure of AI models. we have to look for a solid methodology for constructing and analysing data, we have to create standardised databases, and we have to precede the construction of the algorithms with a serious analysis of the quality of the data. This is painstaking and less rewarding work, but it is necessary, because a poor quality dataset produces poor predictions and in healthcare this has a serious impact on patients.

4.1.3 AI and Holograms

Science fiction has long since detailed the possible applications of holograms. Devotees of the Star Wars Saga remember that already in the first film of the series Princess Leia sends her message for help through a hologram that is stored in the memory of the android C1-P8 (R2-D2 in the original version). Star Trek fans also remember the medical hologram in the Star Trek Voyager series built as a sophisticated mix of artificial intelligence and holographic techniques and which manages to completely replace a real character, also having the advantage of not falling ill during epidemics that it can cure without having to take precautions. Today, we have almost succeeded in making real what was described in Star Wars, what is described in Star Trek, on the other hand, still remains largely science fiction, but can easily become reality when the computing power of quantum computers becomes available. Thinking carefully, one cannot fail to recognise that the idea of combining holographic techniques with artificial intelligence will certainly be the way forward in generating applications and algorithms that can have human characteristics. This may be true for three reasons. The first is related to lower costs. By the time holographic techniques have developed, reproducing a faithful image of a human being will be possible at extremely low cost. Rather than building androids, it will become more useful in many cases to build holograms capable of interacting with the external environment. The second is related to the fact that much of the interaction with the outside world will consist of actions that do not have a physical content, but will consist of a series of digital inputs that are perfectly manageable through interacting holograms. Finally, the degree of fidelity of holographic reconstructions with respect to the original model will be very high, much higher than that of reconstructions made through real androids that will ultimately only be useful to replace humans in heavy labour. After describing what holograms will be in the future and before describing what uses are already real and possible today, it is necessary to explain what a hologram is. Holograms were invented in 1947 by a Hungarian physicist Gabor, who received the Nobel Prize in 1971 for this discovery. Holograms exploit the special property of lasers, which is the coherence of light. A beam of light is said to be coherent when the phase difference remains constant over time. This is the exact opposite of natural light which is extremely incoherent. This phase constancy can be exploited for example in laser surgery to perform micro-operations with a very low degree of invasiveness. Holograms are a combination of two laser beams, the first of which is diffracted by the object, the second is reflected by a mirror, which are made to interfere with each other to construct very faithful 3D images. The progress made by holographic techniques has been enormous in recent years, although we must admit that we are still quite far from what science fiction imagined.

The first holographic films are being experimented with, films that will have a very high degree of realism. Until now, holograms have almost exclusively been printed, but this has obviously limited their potential. Researchers at the University of Tokyo have created holographic films by projecting the holograms onto metasurfaces (surfaces that are only a few nanometers thick) and which allow light to be manipulated. In this case, the researchers constructed 48 frames and, referring to the old Lumiere brothers' techniques, projected them in sequence. What came out was a three-dimensional moving image that appeared as if suspended in air. The still critical point is that to realise a few minutes of projection today requires about a thousand hours of work to produce the frames.

Also in Japan, but at the University of Tsukuba, researchers are working on another innovative project, which is that of interacting holographic projections. This is a very promising field because researchers have managed to produce holograms that change when touched. Using a femtosecond laser, they have succeeded in creating holographic touch images that will soon allow them to extend the potential of augmented reality, even to the point of creating holographic consoles through which one can interact with the environment and which can, for example, be controlled with one's gaze.

Alongside these futuristic applications, however, there are already many real and concrete applications of holographic techniques in medicine.

Diagnostic imaging has had and will have an even greater boost in the future from the use of holographic techniques. Ultrasound, CT and MRI scans with holographic techniques will allow a quantum leap in the possibility of early diagnosis of many diseases. Today, it is already possible through augmented reality viewers to interact with the images produced by diagnostic instruments in 3D format. This means that the doctor will be able to analyse the reports by turning them over and looking at them from several different perspectives and thus having a greater degree of accuracy in the analysis. This technique is used, for example, to better plan complex surgical procedures, e.g. maxillofacial surgery, by having faithful, three-dimensional reconstructions of the entire skull.

Also in surgical practice, holograms will be able to assist the already established techniques of robotic surgery, allowing more precise and less invasive procedures. Thanks to these techniques, the life of a six-year-old girl, Melissa, who was suffering from a rare and serious cancer that had developed on the walls of her heart, was saved. Through the holographic reconstruction of the little girl's heart, it was possible to carry out an operation that preserved the heart, removing only the diseased part with extreme precision. Until recently, the cancer would have been inoperable and the little girl would have faced certain death within a short time.

Further interesting developments may come from combining holographic diagnostic techniques with artificial intelligence algorithms. The National University Health System in Singapore is working on an algorithm that predicts fall risks for patients on wards from analysis of 3D holographic images. The assessment of such risks is crucial to prevent falls and fall-related injuries for hospital patients. Also in the same hospital, a decision-making platform for the management of gastro-oesophageal cancer has been developed that allows sharing of patient data and 3D diagnostic imaging reports to obtain consultations from multidisciplinary teams remotely. This platform is very useful for small peripheral hospitals that can receive rapid consultations and up-to-date treatment recommendations by simply entering clinical data.

Finally from 3D holographic reports, an artificial intelligence algorithm can accurately analyse the back of the eye and accurately detect whether there is an eye condition that indicates a brain cancer or other life-threatening or sight-threatening condition.

An analysis of these examples strongly suggests that there is a need for healthcare to make the leap towards advanced digitisation, both to improve care and the chances of recovery and to reduce costs. The digitisation of healthcare therefore produces a double benefit, and investing in these issues should not only be a strategic objective to reflect on, but first and foremost an almost taken-for-granted fact in the light of a serious cost–benefit assessment.

4.1.4 The New Frontiers of Programmable Biology

The idea behind the applications of mRNA is simply disarming. Until now, medicine, through drugs, has tried to correct protein imbalances that underlie many, if not almost all, diseases. And when this was not possible, the disease was incurable. In reality, our body is capable of producing any type of protein on its own, if given the right input. This process is mediated by mRNA, which gives cells the instructions to build the necessary proteins. Being able, therefore, to create a synthetic mRNA could in principle turn our organism and our cells into a large tailor-made drug factory capable of curing any disease.

The idea is certainly correct and promising, but when it has to be put into practice, it runs into problems, some of which have taken several years to overcome. The Covid 19 pandemic gave a great boost to research in this field, allowing access to large amounts of public funding. The results have been very promising, opening up the field to other important applications of mRNA technology, but at the same time hinting at new problems and also raising ethical questions.

It was in 1990 when the first inoculation of mRNA molecules into an animal was attempted. The experiment was initially successful, in the sense that the organism actually received the instructions conveyed by the mRNA and produced the required protein, but soon afterwards the organism's immune system reacted to the inoculation, identifying it as foreign to the organism, and, as a result, a powerful immune response and inflammation developed. They were thus able to make the organism produce the desired proteins, but at the price of developing a cytokine storm. It was many years before researchers succeeded in designing synthetic mRNA formulations that would fool the immune system and that would transpose suitable synthetic mRNA vectors, lipid nanoparticles, that could also be adjuvants in the process. Covid vaccines based on mRNA technology were the first example of a drug that uses the organism itself to produce the desired proteins. These vaccines were a great success because they showed better performance than other vaccines (e.g. viral vector vaccines) with also, at least as far as known to date, fewer side effects.

The winning feature of these vaccines was also the speed of production, which allowed Moderna and Pfizer to beat the other pharmaceutical companies that were starting from an advantageous position in having an established vaccine platform already available. Moderna produced the experimental vaccine even before the first case of Covid 19 had been identified on US soil! This speed will be a further strong point of this type of vaccine, when it is necessary to adapt the vaccine to variants, which with this technology can be extremely rapid (within a few days) in spite of the longer time (several months) needed to update other vaccines based on viral vector technology, which would also suffer from the problem of the presence of an immune memory in the body in relation to the viral vector.

mRNA technology opens up interesting scenarios for pharmacology. Firstly for vaccines for other diseases. There are plans for mRNA vaccines against malaria, a disease that is very difficult to attack with vaccines because it is accustomed to tricking the immune system, and there is speculation that an effective vaccine against HIV could also be created. It is in principle possible to envisage the creation of universal vaccines for coronaviruses and influenza viruses in the near future, and universal vaccines protecting against a wide range of diseases.

Other applications may be those related to the development of mRNA-based treatments for tumours. What we now generically classify under the heading of cancer disease is actually a very composite and diverse class of pathologies that, although they have aspects in common, are very different, both biologically and in terms of their impact on the body. Modern and futuristic mRNA-based cancer treatments will be very similar to the working mechanism of vaccines. They will start by classifying the neoplasm in order to identify the protein key that enables the body to recognise it and eliminate it by activating the immune defenses. It will also be possible to prevent the onset of particular forms of cancer, particularly those caused by genetic defects. A decisive step in the development of mRNA drugs will be the refinement of the so-called self-amplifying mRNA, which will enable cells to amplify and reproduce the message encoded in the mRNA and thus reduce the amount of mRNA that needs to be inoculated to have any real effect.

A further field of application could be CRISPR gene editing, using mRNA techniques. This could make it possible to switch genes off or on, correcting defects that are at the root of many genetic diseases, but it also opens up ethical questions about the uses of these technologies, which, for example, could also be applied to the attempt to develop modern and sophisticated forms of eugenics. Until now, CRISPR gene editing was a fairly complex procedure and, therefore, more ethically controllable. The development of mRNA techniques for gene editing could allow it to be used on a large scale and at a low cost, with consequent problems of ethical control of the effects of these methodologies.

Using it, for example, to switch off the sickle-cell anemia gene that still causes thousands of deaths in Africa today would be progress for mankind, developing even subtle forms of eugenics would, on the contrary, be a big step backwards.

Leaving aside the ethical aspects, more generally, one question that needs to be answered is that of the limits of applicability of this technology. It would be too optimistic to think that it would be possible to use mRNA drugs for all diseases. There are likely to be classes of diseases for which these techniques will be decisive and others for which they will be of no help. The mRNA technique is not the panacea for every disease! Liver diseases will probably be an easy target for these technologies, because lipid nanoparticles are attracted to the liver, while they will probably be ineffective for other diseases. Recognising these limitations is the first step to making important pharmacological innovations. mRNA vaccines have been a success and were mankind's first step towards programmable forms of biology, which are very promising from a medical and scientific point of view, but on which much remains to be investigated and many ethical questions remain to be answered.

Protein folding has always been a problem considered almost irresolvable in biology and biochemistry. Some (Chemistry World) even went so far as to consider it the Holy Grail of Chemistry.

It was announced recently by Deep Mind, a company that deals with advanced applications of artificial intelligence, that it had succeeded in solving this problem through an artificial intelligence algorithm, Alpha Fold. This development, according to some, would have advanced mankind by an estimated two to three decades, and has obviously caused much interest and excitement among researchers around the world. This is an achievement that has the potential to be a breakthrough for biology and biochemistry comparable to that of penicillin and antibiotics could revolutionise many areas related to biology and biochemistry.

This discovery, made through the simulation of 3D models of proteins, could, for example, open up extraordinary possibilities in terms of the development of new disease-specific drugs and also innovative methods for waste disposal.

The Alpha Fold deep learning algorithm is able to predict new protein structures and their evolution at the atomic level from a database of 170,000 known protein structures within a few hours with a precision that can now be achieved with sophisticated crystallographic analyses. Until now, the crystallographic study of a protein required many years of work by a team of researchers.

This is undoubtedly a great achievement that can open up great developments and be a harbinger of numerous applications. Let us not forget that diseases resulting from protein errors are gradually increasing their incidence due to a whole series of factors. Alzheimer's is perhaps the most widespread and best known, but there is a whole category of diseases that in technical terms are called protein misfolding diseases. These are diseases for which the prognosis today is generally inauspicious and for which there are no cures or only symptomatic treatments that can only delay but not cure the disease. The common feature of these diseases is that altered proteins lose their original function and bind with other proteins causing cell death. Alzheimer's belongs to a group of so-called neurodegenerative diseases that have in common the formation of amyloidosis leading to the death of neurons. Parkinson's disease and Creutzfeld Jakob syndrome (also known as the human variant of mad cow disease) also belong to this category.

Other diseases are related to protein transport and are Diabetes insipidus, Gaucher's disease, Retinitis pigmentosa, Fabry's disease. In addition, many cancers are caused by protein misfolding.

To detect a single folding error until now required several years of study and this is the main obstacle that has limited the development of new drugs for these diseases. The possibility, therefore, of being able to obtain three-dimensional models of altered proteins in a few hours is a huge step forward in drug design for these very serious diseases.

Alpha Fold through deep learning is able to do in a few hours the work that a team of biologists and biochemists would do in three to four years, and moreover, it can do it for a large number of proteins, thus reducing errors. And the roads to finding proteins capable of targeting altered proteins would be all downhill.

Obviously, Deep Mind's announcement alongside a part of truth also has a propagandistic content. There is a first problem, if you like of an epistemological nature, which consists in having to accept a prediction of protein structure without having experimental verification available. That is, we can trust 100 per cent what the algorithm says, even with regard to the degree of accuracy required. Would it be conceivable to try to make drugs from a simulation of protein misfolding?

Furthermore, we are dealing with a simulation of protein structure, but we are not, with this methodology, able to say anything about the rules governing protein folding. According to some, in fact, Alpha Fold would appear to simply evade the problem of protein folding, because we can only come up with a final model, without being able to infer anything about the rules of evolution. However, the development of specific drugs cannot, in principle, be based on the folded protein model alone, because research is discovering that the shape of proteins is only one piece of the reasoning as it can lead to non-unambiguous results because proteins with almost identical structures can behave chemically very differently and very different structures can behave similarly.

Therefore, while Alpha Fold's result is certainly interesting and innovative, its impact on the possibility of producing specific drugs for protein misfolding diseases in the near future must be downplayed. Artificial intelligence has provided us with one more tool to operate in the field of these diseases, but the creation of effective drugs is still a long way off. This is actually a significant step forward, but probably not as revolutionary as some might have intended. Artificial intelligence has only opened up new horizons for us, but there is still a long way to go.

4.2 AI, Economics and Ethics

4.2.1 AI, Antritrust and Competition Between Large Economic Groups

FAANG is a widely used and very current acronym in the Western world. It stands for Facebook, Apple, Amazon, Netflix and Google. This can be contrasted with the acronym BAT **(Baidu, Alibaba and Tencent)** in China. These acronyms are monopolising the market and governing our lives as well. And in the coming years, one can assume that the confrontation between FAANG and BAT will be one of the most important battlegrounds on which the geopolitical and economic competition between the West and China will be played out. Before the pandemic, it was predicted that BAT would soon catch up with FAANG in terms of turnover with even greater growth prospects due to the greater growth potential of the digital sector in China than in the West. The pandemic has shuffled the cards a bit, but a physiological decrease in consumption due to the global recession has to be contrasted with a higher rate of utilisation of the digital economy and transactions driven by the pandemic and lockdowns, which can be a strong point for the digital giants in the future.

FAANG's turnover in 2018 (with the addition of Microsoft) was about $700 billion, almost as much as Switzerland's GDP and equal to 0.5% of world GDP. BAT was expected to reach the same level of turnover in 2020 with a potential market of $2 trillion and higher growth trends. According to Italian Agcom's 2019 Digital Platforms Observatory, platforms (the figure refers to FAANG, but is easily extendable to BAT as well) have "a gross profitability of 49% and an operating margin of 21% of revenues. The return on equity (32% on average over the past three years) and on invested capital (15% average ROI over the past three years) are very high. Values decidedly higher than those found for the main TLC&Media companies and for the more than 2000 largest Italian companies'. Again according to this same report, the platforms that "globally occupy the top positions in all the business sectors in which they operate, and the market shares of the companies that rank first are never less than 30%". Very high—again according to this report—are the minimum turnover sizes in e-commerce to make a profit, which are over 50 billion euros, while a search engine has a breakeven around 20 billion.

These figures describe a highly concentrated market with high barriers to entry, and with these numbers any industrial sector would be far from competitive. In the case of the digital economy, the market power of firms operating in it is not only limited to aspects of market share or business profitability as would be the case in traditional industries. In the digital economy, market power can also manifest itself through other variables that make it even more problematic to ensure competitive conditions.

In such a scenario, aggressive forms of collusion or concentration could grow out of all proportion, distorting the proper competitive functioning of markets.

A first form of collusion consists in using artificial intelligence algorithms that have the same architecture and database. In this case, it is all too obvious that there will be equal answers to every request. The banking system today, for example, bases its verification of customers' creditworthiness on artificial intelligence algorithms. If the architecture of the algorithm is the same and the big data referred to is the same, all banks will grant the same conditions to each customer. If the bar is set high, the credit market is rationed, increasing profits for banks and decreasing credit potential for consumers.

A second form of distortion of competition is that which can occur through the processing of data derived from user profiling with the possibility of perfect price discrimination. This operation grants enormous market power to those who are able to control the profiling data with the consequence of an easy elimination of competitors. It could even lead to forms of algorithmic collusion that manipulate information to move up the market price bar. Moreover, market power in the digital economy does not depend on market share, but on the amount of data possessed and the processing and calculation capabilities. A company that concentrates large amounts of data and large processing and computing capacities is able to exert market power, distorting competition. This means that new mergers must not only be assessed on the basis of turnover, but also and above all on the amount of data possessed and the resources in terms of processing and calculation capacities.

These considerations make us understand why in the space of just a few years the Tech Giants have been able to grab significant and ever larger shares of the world's GDP by using boundless monopoly power to increase their profit margins to the detriment of both businesses and consumers. The future competitive scenario will see large agglomerations of digital companies from certain geographic and geopolitical areas confront each other (BAT vs. FAANG), but always within a market that will have very low levels of competitiveness. If you like, this confrontation will be one of the most important areas of a global competition between the West and China that, before being military, will be economic and will see the telecommunications and digital platform sectors as a privileged area of conflict. Let us not forget, then, that digital platforms have a great asset, which is the possession of boundless databases of personal data and profiling. These data lend themselves not only to licit commercial uses, but also to illicit and improper uses if they are used to manipulate information and the proper functioning of democratic institutions. And this power in a scenario of global competition between the West and China is not unlikely to be used on a large scale in the coming years. Antitrust policies thus become not only an instrument to protect the market and consumers, but also democracy and civil liberties.

The large-scale use of machine learning algorithms linked to the availability of big data opens up boundless frontiers to new forms of competition, some of which are lawful, others trespassing dangerously towards monopolistic or oligopolistic forms of market management.

In such a scenario, aggressive forms of collusion or concentration could grow out of all proportion, distorting the proper competitive functioning of markets.

The market power of the so-called Tech Titans can be expressed in three different forms:

(a) predatory prices;
(b) limiting competition from new entrants;
(c) creation of opportunities in other markets.

Predatory pricing is a tool to drive competitors out of the market so that they can enjoy the benefits of operating under monopoly conditions in the future.

The limitation of competition from new entrants in terms of both companies and products consists of all those policies aimed at raising barriers to entry into certain markets for products and companies; the creation of opportunities in other markets consists of using the market power gained in one sector to extend it to neighbouring sectors.

A prime example of problems with the market power of Tech Titans is the so-called Amazon case in Europe.

At the moment, the European Union represents one of the most strenuous bastions of competition defense, considering that the United States, which historically was the home of antitrust policies with the famous Sherman Act of 1890, has gradually weakened its margins of intervention in the markets, also as a result of numerous Supreme Court rulings that have greatly restricted the scope of the Sherman Act.

The area that European regulators are investigating concerns Amazon's treatment of third-party sellers and the conflict of interest that arises from being both a seller of products and a platform for the sale of the same products by other sellers. In essence, Amazon exploits its market power to steer the sale of its products to the detriment of third-party sellers, in that when it captures the market's interest in a certain type of product, it offers them at lower prices on the market, damaging third-party sellers.

This kind of unfair competition would have to do with predatory pricing.

The problem of unfair competition from Tech Titans who are also gatekeepers has also spread to other platforms. Apple is also under investigation by the European Antitrust Authority for the Apple store platform in relation to its relationship with developers. Apple in its commercial agreements imposes two restrictions, the first consisting in the prohibition to provide information to users on how to access Apps outside the proprietary platform and the obligation for developers to pay a fee for licenses sold through the Apple store.

This would be the exploitation of market power to limit competition from new entrants.

It should be pointed out that already in 2018, the European antitrust authority had sanctioned Google, because it had forced Android users to route searches through its search engine to the detriment of other competitors.

The traditional concept of concentration was based on market share. This concept appears inadequate for the study of what we might call the 4.0 trusts. Market power in the presence of big data and artificial intelligence does not depend on market share, but on the amount of data possessed and the processing and calculation capabilities. A company that concentrates large amounts of data and large processing and computing capacities is able to exert market power and distort competition. This means that new mergers must not only be evaluated on the basis of turnover, but also and above all on the amount of data held and the resources in terms of processing and calculation capacities. Recent acquisitions and business deals should also be seen in this light. For example, the *Facebook/WhatsUp*, *Microsoft/Linkedin* combinations are worth much more in terms of market power than the mere sum of turnovers.

Three conclusions can be drawn from these considerations:

(1) Do we need new instruments of control and new antitrust policies? The answer is no. Traditional antitrust policy instruments work perfectly well in the 4.0 scenario, even if they have to act on different variables than previously considered. They must consider information and processing capacity as the origin and source of market power and consider aggregations between companies not only in relation to turnover.

(2) Can antitrust policies be delegated the role of defending the consumer from privacy violations, cybersecurity breaches and information distortion?

    Again, the answer is no. Antitrust policies should only act when there is a real distortion of competition and harm to the consumer, as the mere acquisition of a dominant role by a company is not sufficient, which, in the final analysis, is precisely the goal of healthy competition between companies. Furthermore, the use of antitrust policies cannot be seen as a substitute instrument for other sensitive fields, where the instruments of protection must be different.

(3) Not the tools change, but the way they are applied and the mix that must be used for proper regulation changes. Reshaping antitrust policies in the presence of big data and artificial intelligence is one of the most important economic challenges of the coming years.

The technology conflict between the US and China began when US industries first perceived that they had lost their monopoly on technology innovation in emerging high-tech sectors. Throughout the twentieth century, the US was the technology leader in almost all sectors. China has slowly closed the gap and in recent years has begun to undermine the US in many areas, first and foremost in telecommunication technology and artificial intelligence. *Last but not least,* in this scenario can be placed the technological competition for the discovery of a vaccine against Covid 19, which has seen China jump the gun and be the first, in parallel with Russia, to register a vaccine, albeit with very poor testing procedures that are potentially dangerous for the safety of the populations that should be subjected to mass vaccinations. The war in Ukraine and the subsequent radical change in the geo-political scenario have made this conflict, which still burns under the ashes, even more creeping and hidden. Today, it is clear to everyone that the dominance of technology represents a very powerful weapon to be used to turn the planet's delicate geo-political balance in one's favour. It is a kind of underground war, apparently bloodless, but conducted with the same rules as real war. The end of Trump's presidential term has sweetened, at least apparently, the trade war for the control of technology that was being played out between the United States and China and that had as its first act, the ban imposed on Google from supplying technology and therefore software to Huawei, the Chinese telecommunications giant that has challenged Samsung's dominance in the field of smartphones and is becoming the most important technology provider for the management of 5G networks. Against this backdrop, one can understand the political significance of Google's ban on supplying Huawei technology and the substantial ostracism that all Western countries have put in place by effectively excluding Huawei from the ranks of those who can manage each country's domestic 5G networks.

Another chapter in this undeclared war was the failed attempt to ban the downloading of the Tik Tok app on US territory. The blocking of Tik Tok was supposed to be triggered, after some postponement on 27 September 2020, but was blocked by a judge the following day and subsequently dropped. If one takes into account the agreement between Tik Tok and Oracle, it is easy to identify the reasons why the blockade was not implemented This app has been a great success, especially among young people and has been downloaded by more than 100 million people, undermining the market share and power of other apps such as Facebook, Whatsapp, and Instagram. The Tik Tok app has as its strong point a sophisticated artificial intelligence algorithm that suggests the next video to watch to users based on their preferences. It is quite easy to understand how this is the key to Tik Tok's success, but also one of the sources of concern for governments, because the app with this function has the possibility of directing the behaviour, choices and decisions of users, even going as far as forms of covert mass persuasion, or conditioning of electoral mechanisms in Western countries. In addition, the Chinese operator has access to the personal data of more than 100 million Americans. To overcome the impasse, ByteDance, the owner of Tik Tok, entered into commercial negotiations to find an American partnership to operate the app on American soil. After an initial approach with Microsoft, which failed because ByteDance refused to include in the contract the sharing of the artificial intelligence algorithm at the heart of the Tik Tok system, ByteDance reached an agreement with Oracle. This agreement gave rise to a new company under US law that is to manage the app's data flows, in which ByteDance has a majority share and Oracle a minority share. In essence, Oracle becomes a technology partner for the provision of the hardware and software infrastructure for the management of the app in the US. But it does not have access to the artificial intelligence algorithm. It has, therefore, the possibility of controlling the source of the app, preventing doors from being opened that would lead to the transmission of users' personal data, but it does not have access to the app's real asset, which is its algorithm. This solution solves the doubts about the security of American users' data, but does not solve the problem of the potential social interference that the app could exert if the artificial intelligence algorithm were geared towards this purpose.

However, the benefits for Oracle are also considerable. Firstly, it strengthens its market share in the Biz Cloud field, increasing its turnover by USD 100 million and establishing itself as a first-class technology partner. The newly created company also generates at least 25,000 new jobs. ByteDance has the advantage of retaining control of the algorithm that is the company's most important treasure and can with this agreement overcome the restrictions imposed by the US administration. From a commercial point of view, this marriage turns out to be a bargain for the two parties, however, it leaves doubts as to the conditioning possibilities inherent in the app, which in this case remain under total Chinese control. Certainly the good relations between Oracle and the Trump administration have facilitated a process and prevented anyone from going so far as to say that this marriage is a no-go. The loser in this battle is Microsoft, which sees two of its competitors strengthening within its own core market. However, beyond the economic benefits for the contracting parties, this affair marks a point in favour of China in the broader war for technological supremacy with the United States, and only the next few years will let us understand how much this can weigh on the world geopolitical scenario.

4.3 AI, Algorithms and Work 4.0

Charlie Chaplin's famous film 'Modern Times' marks a very strong criticism of the organisational paradigm prevailing at the time, known as Taylorism, which consisted of an attempt to standardise processes, treating workers like machines with the aim of increasing labour productivity. Over time, labour legislation has found ways and means to protect and defend the dignity of workers from the excesses of Taylorism. Industry 4.0 and artificial intelligence are now elements of the production process that make some of the instruments of labour protection obsolete and require urgent updating.

The protection of the dignity of work and substantial respect for workers' rights become objectives to be protected by new instruments. It is human labour as a whole that therefore needs new instruments of protection.

In this process of identifying new instruments of labour and worker protection, the first aspect to consider is the initial selection of the worker.

The use of artificial intelligence tools in selection poses several protection problems.

Entrusting the choice of the best profile to an algorithm may be a quick and inexpensive way to select, but in the first place it exposes the selection to the possibility of the artificial intelligence developing biases that may invalidate the fairness of the process. The other important aspect is the transparency of the procedure, that is, the right of the evaluated person to know both the information base and the algorithm on the basis of which individuals are evaluated.

Another interesting aspect is the improvement of production processes achieved by monitoring work activities. A general principle is that control of the human being by a machine or remote control system is not permissible. However, in Work Organisation 4.0, it is almost impossible to separate process monitoring from worker control. Wearable devices, smartphones or other process monitoring tools collect a whole range of data not only on the productivity of the process, but also on the productivity of the worker and his behaviour during the production process. They could also in principle be used to obtain information on the health of the worker. Control systems in a certain sense become part of the production process and the difference lies only in the type of data that is collected and processed. The protection of the worker, his privacy and the prohibition of remote control then become almost impossible because it is impossible to discriminate between the production system and the control system.

It is therefore clear that the instruments of protection need to be innovated and refined. Probably the concept of work itself has to be changed because the increasing interconnection between human and machine work makes it difficult to distinguish the individual contribution.

The so-called GIG economy is an economy in which the employment relationship is reorganised to serve the digital economy. The algorithm may not just monitor the worker, but may be turned into a manager and instructed to organise work, using the large amount of data it can collect. The same algorithm could also easily develop discriminatory biases and tend to exploit or marginalise certain workers.

The algorithm manager can certainly make the production process more efficient, but it can easily give rise to a new Taylorism 4.0.

The risk of exploitation and the risk of worker discrimination appears significantly higher.

This scenario also entails the need to review the instruments for protecting workers. The definition of specific legislation on Work 4.0 is beginning to show itself necessary and this need will become more and more compelling in the coming years.

It is therefore necessary to rethink trade union freedoms and collective bargaining, the fight against direct and indirect discrimination, the fight against forms of undeclared work and the new regulation of informal work, the design of tools to control working hours and to avoid forms of 'forced labour' within the framework of a 4.0 labour legislation. We are in danger of falling into a kind of 'algorithmic caporalism' and to defend ourselves against this the regulations currently in force are not very effective. Today, it is possible with apps to contract a work performance of a few minutes, in ways that ensure a downward competition of the remuneration of that work and, above all, without any contractual or security constraints. On the one hand, demand can be parcelled out and the performance underpaid; on the other hand, there is an interest in performing as many services as possible, neglecting all forms of labour protection. The algorithm that oversees this systematic process of labour exploitation and ensures its economic efficiency is nothing more than a digital corporator that treats labour as a commodity and attempts to make the maximum profit in spite of all protective regulations. Writing new regulations for Work 4.0 to make it capable of governing the digital and gig economy is certainly a difficult and challenging, but necessary goal.

4.4 AI, Poverty and Inequality

In many countries of the world, digital mechanisms are being explored to manage the welfare system in a way that makes it, at least in intentions, more transparent, fairer, cheaper and more efficient.

Transferring all benefit-related requests for assistance onto a digital platform and replacing the human factor with an algorithm in the management and processing of the file may seem like a very good idea and a great way of exploiting new technologies to the benefit of the weak. Indeed, some of the more optimistic observers might even entertain the idea that this is the main road to the relatively quick abolition of poverty. Unfortunately, things are not exactly like that, and the reasoning we are going to develop may lead to interesting considerations on the value and risks of artificial intelligence.

A first consideration that needs to be made when reflecting on the risks of such a mechanism is to ascertain when the 'digital divide' between the poor and the rich and within the category of the poor, between different ethnic groups and different types of poverty, can affect the efficiency and effectiveness of the digital welfare system. The poor are, of course, those with the least capacity for digital access and this 'digital divide' grows the more severe the poverty becomes. A first criticism that can be made of the digital welfare system is that it definitely increases inequality because the poor have less capacity for digital access and among the poor, capacity for access is inversely proportional to the intensity of poverty. The poorer one is, the less one is able to access digital resources, with access capacity being understood both as the availability of appropriate hardware and software and the ability to access and make effective use of IT resources in terms of knowledge. In this sense, digital welfare undoubtedly penalises the poorest and the least guaranteed, operating in a manner diametrically opposed to what should be the logic and philosophy of the welfare state.

A second point that must be taken into account is that the lack of a human interface makes it difficult and complicated to solve simple problems that a human with common sense can bring to a successful conclusion in a short time and that instead become insurmountable obstacles for the algorithm. And since it is on the proper functioning of the welfare system that the often necessary and fundamental assistance of individuals is based, the limitations of the algorithm can be devastating to the lives of individuals who are already in a precarious situation. Consider that for a person enrolled in a food assistance program, losing his or her digital identity credentials can mean having his or her subsidy cut off, leaving him or her without assistance and risking starvation. Think of the devastating effects of software that fails to complete the procedure in special cases, condemning the subject to not being able to access assistance, or more trivially the effects of the subject's inability to close a complex procedure, which can lead to the suspension of assistance. If the problem-solving interface is human, one almost always succeeds in overcoming these bottlenecks; if the interface is digital, then the most frequent response will be: 'it is not possible to conclude the procedure', with enormous damage to the life of the sufferer.

A third and final point that highlights the risks and criticalities of the use of algorithms in the management of welfare systems is related to the possibility that the artificial intelligences that govern the assessment processes related to digital welfare may develop biases that penalise or favour certain individuals in relation to gender, race, education, religion or political orientation.

These considerations lead to the inference that a digital welfare system may increase social inequalities, widening the gap between the rich and the poor and increasing the gap between the least poor and the poorest.

By analysing concrete cases, we can better understand the contours of this discrimination.

The idea of catching the fraudsters who receive the benefit without qualifying for it is ingrained and ingrained with the development of welfare systems. The thing that until now has discouraged large campaigns against this kind of illegality is the high cost of control in relation to the benefit that can be obtained. If I spend more to catch a scoundrel than I recover, the game is obviously not worth the candle. With the use of algorithms and big data, the cost of the search becomes considerably lower and more acceptable, which is why many administrations have made extensive use of these tools in different parts of the world. In the state of Indiana, a large campaign was launched to identify welfare recipients who were not entitled to it, while in Australia, large campaigns were launched to recover overpaid subsidies. But paradoxically in this case it is precisely the extreme power of the algorithm that is a point of weakness. It is an attempt to bring illegality to light, but it is too much like trawling and has significant side effects. The search process begins with the selection of a group of potential fraudsters that is then subjected to further scrutiny. But the risk that in this group of potential fraudsters there are subjects who are completely in the clear is quite high, and the subsequent stages of detecting illegal situations through telephone interviews, far from reaching the possible illegal recipients, indiscriminately target the weakest subjects, those who have the most difficulty with a telephone interview or providing documentation via the web. In the end, the outcome of the process is that a large number of individuals, only a fraction of whom were not actually entitled to it, are de facto deprived of the subsidy, while in most cases the subsidy is taken away from people in need who are left, as a result of this injustice, severely penalised and without social protection.

Ultimately, then, the use of these technologies severely restricts the right to privacy of the poorest individuals who are forced to share their personal data, and in some cases even sensitive data, in order to allow the algorithms to train them without being able to oppose any denial, because not providing these data means certain loss of the subsidy. From the condition of poverty thus arises the impossibility of fully exercising the right to privacy.

Thus, digitising the welfare system and using algorithms instead of a human interface appears to be an inefficient and unfair solution, which not only does not improve the quality of life of people involved in welfare programs, but can also lead to dangerous situations for those enrolled in these programs. It can be said that such a model penalises the poor by making them even poorer. Certainly, the idea of eliminating the empathy linked to the human factor apparently makes anti-poverty systems more efficient, but in reality it only makes them lose the content of humanity that perhaps constitutes an essential and irreplaceable element for the success of the process.

Fighting poverty is not easy, simple recipes or simple incentives are not enough, and subcontracting the conduct of this complex and difficult war entirely to machines and algorithms can be a very bad idea!

4.5 AI, Robotics and Ethics

The android units D-3BO and R2-D2 are characters from the Star Wars sci-fi universe that have fascinated the minds of many devotees of the series. Today, however, what could have been imagined as science fiction applications a few years ago are now almost completely part of reality and can make a very important contribution to us at a time when the Covid-19 forces us into social distancing and makes it necessary in some cases to limit our relationships with humans.

Having a robot as an aid for our daily activities is now a frontier reached!

In general, one can think of different levels of involvement of robots in interaction with humans. Each level corresponds to a certain level of task complexity.

The first and lowest level consists of the use of robots that perform simple tasks by replacing human intervention. This use can be extremely useful and economical in certain situations. A robot that acts as a doorman is an example of a cost-reducing tool, while a robot that sanitises by entering a biohazard area is not only a cost-reducing tool, but also a tool to reduce the risk of exposure for workers.

A second level of applications consists of the use of robots that perform complex tasks and provide an intelligent interface for the remote management of activities.

The first use we instinctively think of here is that of robots in healthcare.

The creation of a hospital with robot nurses and tele-medical care of non-critical patients is an extremely useful technological response to disease in the event of disasters or epidemics. In the case of Covid, the use of medical robots made it possible to decrease contagion among healthcare workers and to focus care on the critically ill. The use of particular types of robots that are also capable of video calling can be a useful support for patients, allowing them to stay in contact with family members. Moreover, this capability, implemented in the form of video healthcare calling, constitutes an extraordinary tool for diagnosis and treatment that on the one hand allows 118 operators, for example in the event of an accident, to have a real time picture of the patient's condition and thus be able to prepare the most appropriate means of intervention. It allows the realisation of a preventive triage that can better redirect the entire subsequent care pathway. But video healthcare calling is also very useful for territorial healthcare because it reduces travel time, can be carried out over a very wide time slot, and creates a direct relationship with the patient. In general, it will not be the patient who has to move, but it will be the doctor who, through the telematic tool, will reach him or her at home, also having at his or her disposal the monitoring data realised through the connected devices. For frail or bedridden patients, this is a great help, but this tool can be an extraordinary means during the management of the Covid 19 epidemic to control compliance with the quarantine and to monitor the condition of positive patients in home care, without putting healthcare personnel in danger of becoming infected.

Other examples of robots capable of performing complex tasks include the Sky Sports robot that effectively replaces a television operator, a robot that delivers food and receives payments, a robot waiter that makes coffee or serves food and drinks while also taking orders.

Also belonging to this category are robots that can be used for teaching or information purposes and robots that can distinguish particular situations such as social distancing and the use of masks.

However, the real quantum leap lies in the third level of robot utilisation where the AI-equipped robot does not just perform predefined tasks or merely act as an intelligent interface for humans, but is able to make autonomous decisions.

This third level of applications is the advanced field of research in which artificial intelligence scholars are working. The android Sofia is a robot capable of perceiving the external environment and developing autonomous behaviour in relation to it. Robots equipped with 'emotional' software could be used to keep the elderly or children company, although recent experiments have shown that such emotional software can develop biases that can lead them to be dangerous, especially if they enter into relationships with fragile or easily influenced subjects.

Another example is Asimov, a robot that can act as a multi-task mobile assistant and can be thought of as the big brother of the virtual assistants that are now quite common in the home.

Also famous are the four androids built by Iroshi Ishiguro that are able to give lectures instead of their builder.

Beyond this frontier, science blurs into science fiction and one can imagine building robots in the future that come ever closer to human behaviour. Some even go so far as to hypothesise building robots guided by artificial super-intelligences whose behaviour is indistinguishable from that of humans and who also develop a conscience and the ability to feel, although it is legitimate to harbour many doubts as to whether this will actually become reality.

The dream of a sentient android has inspired many science fiction writers, yet today we are just at the beginning of a process in which mankind will be confronted with new problems and may even be led along the way to having to reconsider its role in the world and the universe.

Boston Dynamics is an engineering and robotics company founded in 1992 as a spin-off from the Massachusetts Institute of Technology that became famous for its robot dog, latterly nicknamed Digidog and first realised in 2005. In 2013, the company was bought by Google, which subsequently sold it in 2017 to Japan's Softbank, which, in turn, sold it for almost a billion dollars to Hyundai. The main asset of this company, which is valued at nine figures, is Digidog, a robot dog costing just over $70,000 per unit, weighing around 30 kg, capable of developing a speed of almost 5 km per hour and able to move autonomously within a radius of 30 m, relying on a whole series of sensors and cameras that make it very useful for monitoring. It can also be fitted with a mechanical arm that enables it to grasp objects and lift weights. Coupling it with the most advanced mobile data transfer systems (6G) can greatly enhance its performance and range, making it de facto completely autonomous and independent, and therefore with a very extended range of action. It is an extremely versatile tool that can be used for very diverse and heterogeneous purposes in the medical field, home monitoring and surveillance, police operations, and warehouse handling. The latest applications that are being experimented with also include the possibility of joint use for a single task of a large number of these robots, which in this case act in all respects synergistically as a pack and can perform complex collective tasks.

As can be deduced from these initial descriptions, it is a tool that can be very useful and versatile, but one that raises legitimate doubts as to whether it can be used as an aggressive and invasive surveillance tool, and even as a weapon system.

The New York Police Department is testing Digidog for use in the place of officers in dangerous situations, for instance in the case of an explosion risk or to enter hazardous environments, but more generally it can be a useful tool to support patrols and territorial control for the police force, especially in very large contexts, for instance in parks.

The first risk that can be attached to this robot is that of becoming too invasive for privacy. Its sensors, microphones and cameras can capture and store much more data than a human can, even if equipped with advanced technology, and can do so dynamically, following even in the steepest and most inaccessible terrain the subjects that need to be monitored. It can be used to monitor parks and sparsely frequented areas of the city that are normally used as drug dens, but the volume of personal and sensitive data it can collect in these operations is enormously large and can pose a danger to the privacy of many individuals, especially if this data falls into the wrong hands.

Think what an extraordinary and terrible tool for repressing demonstrations could be represented by a pack of these robot dogs, which can monitor thousands of people taking part in a procession or demonstration and then, through facial recognition technologies, go and identify them during their daily activities. These robots could also be used to break up demonstrations, teaching them to charge assemblies and marches. Totalitarian regimes would have at their disposal an efficient and extremely low-cost tool to control and suppress dissent, apparently in a less violent manner.

The most serious risk, however, is that these robot dogs could be equipped with weapon systems or more simply with offensive systems (e.g. tasers that do not kill, but stun), posing major ethical problems. In fact, these are not remote-controlled automatons, but machines capable of making autonomous decisions and which, therefore, could potentially even go so far as to kill if they thought that was the best choice, or which could develop racist bias and selectively target certain ethnic groups. What we are describing here are not dystopian scenarios distant in time, but facts that may be extremely relevant today, since it would be enough to replace the mechanical arm with a weapon or offensive system to turn the robot dog into a perfect instrument of death. Of course, evolved democracies could not tolerate such instruments to control individual freedom, but there are still too many totalitarian regimes in the world and too many governments that do not have too many qualms about implementing violent actions to repress dissidents or opposition. The current case of Burma is but the latest example of the destructive potential of these technologies when put at the service of regimes that have no regard for respect for freedom. This is also facilitated by a cost that is all in all quite affordable and a technology that is well established and easy to use. For as little as EUR 1 million, any regime could equip itself with a pack of these robot dogs and then use them as an extremely efficient tool of repression or as an extremely invasive instrument of espionage on individual privacy.

Therefore, the importance of emphasising the need to give ethics to artificial intelligence and to ensure that unethical applications are blocked, also with respect to these tools, comes up again. Technology is neutral; it is neither good nor bad. It is the use that is made of the technology that makes it good or bad. It is, therefore, necessary that experimentation with these technologies that may be invasive to individual freedoms take place in a context of extended artificial intelligence, i.e. in a context in which the human being retains control over decisions, supervises decisions, and is able to intervene and block the machine when, for whatever reason, it oversteps certain limits. But this is, of course, only valid for democratic governments that are supposed to punish certain repressive behaviour, which is different from the case of regimes that have no qualms about implementing harsh repressions and might appreciate the very unethicality of these technologies. Alessandra Ocasio Cortes, a member of the U.S. House of Representatives, has often in the past denounced the racist potential of these technologies used by the NYPD, lamenting the fact that while these new ground-based drones are deployed to police low-income communities of colour, they neglect to provide the most basic technological equipment for schools and healthcare.

Digidog makes us realise how high the level of attention must be with respect to the misuse of new artificial intelligence technologies!

Conclusions

Sit Finis Libri Non Finis Quaerendi

Artificial intelligence and its applications are a fascinating and intriguing topic of study. Many and important are the potential benefits that the large-scale application of these technologies will bring to mankind [2, 3]. However, as with any technology, there is a flip side to the coin, a dark side that needs to be investigated and known even if only to be able to avoid the negative spin-offs. With this work, an attempt has been made to understand what the world we will find ourselves living in a few years from now might be like, a world that will not only belong to our children, but will also encompass a substantial part of our lives. Technology, we know, is neutral in the sense that it is neither good nor bad. It is the user who makes it good or bad. Thus the applications of artificial intelligence are not a priori good or bad, it is the users who can, through them, improve the lives of human beings, but also lay the foundations for the destruction of humanity. Understanding artificial intelligence means understanding the world in which we will live in the near future, in the next 5–10 years.

An attempt has been made to zoom in on certain fields of application of artificial intelligence, without claiming to be exhaustive. An attempt has been made to make, to use cinematographic language, not a systematic film, but a collection of short films capable of enabling the reader to frame the complex world of artificial intelligence at 360°, in a concise manner. As has already happened other times in history, humanity is confronted with an extraordinary technology and the bet that this generation is making is to be able to govern it in order to produce greater well-being for the world [1]. We do not know whether this will be the panacea for mankind's ills, whether it will constitute the recapture of the Garden of Eden where mankind will be free from poverty and the need to work, or whether, as predicted in dystopian scenarios, it will become a nightmare in which mankind will be subjugated and perhaps destroyed by new intelligent artificial entities that will take possession of the earth. Fear of progress has always characterised epochal turning points and at this juncture, too,

D. Marino et al., *Ethics and Artificial Intelligence*, Studies in Systems, Decision and Control 519, https://doi.org/10.1007/978-3-031-50902-5

D. Marino et al., *Ethics and Artificial Intelligence*, Studies in Systems, Decision and Control 519, https://doi.org/10.1007/978-3-031-50902-5

the dark side of this technology is disturbing the dreams of mankind. We are probably (re)discovering that if we do not give ethical content to technology, it may become an instrument of self-destruction, but at the same time this attempt to fill technology with values may turn out to be a huge step forward for mankind. An attempt has been made to describe some future scenarios based on current applications, with the conviction that in every laborious process of innovative research, what seems to be the end of one path, only opens up new horizons and poses new questions, thus becoming the beginning of a new path of research.

Bibliography

1. Dignum, V.: Responsibility and artificial intelligence. The Oxford Handbook of Ethics of AI 4698 (2020)
2. Marino, D., Gil Lafuente, J., Tebala, D.: Innovations and development of artificial intelligence in Europe: some empirical evidences. Euro. J. Manag. Bus. Econ. vol ahead-of-print No (2023). Ahead-of-print https://doi.org/10.1108/EJMBE-03-2023-0085
3. Tebala, D., Marino, D.: Companies and artificial intelligence: an example of clustering with orange. In: Marino, D., Monaca, M.A. (eds.) Innovations and Economic and Social Changes due to Artificial Intelligence: The State of the Art. Studies in Systems, Decision and Control, vol. 222. Springer, Cham (2023). https://doi.org/10.1007/978-3-031-33461-0_1

Sitography

5. http://www.jstor.org/stable/4129589
6. https://ai-ethics.com/2017/08/11/future-of-life-institute-2017-asilomar-conference/
7. https://libro-bianco-ia.readthedocs.io/it/latest/
8. https://en.unesco.org/artificial-intelligence/ethics
9. https://ec.europa.eu/futurium/en/ai-alliance-consultation.1.html
10. https://ec.europa.eu/info/sites/default/files/commission-white-paper-artificial-intelligence-feb 2020_en.pdf
11. https://eur-lex.europa.eu/legal-content/IT/TXT/?uri=CELEX%3A52021PC0206
12. https://ec.europa.eu/info/files/proposal-directive-adapting-non-contractual-civil-liability-rules-artificial-intelligence_en

D. Marino et al., *Ethics and Artificial Intelligence*, Studies in Systems, Decision and Control 519, https://doi.org/10.1007/978-3-031-50902-5